IMAGE AND VIDEO COMPRESSION

Fundamentals, Techniques, and Applications

IMAGE AND VIDEO COMPRESSION

Fundamentals, Techniques, and Applications

Madhuri A. Joshi, Ph.D.

Professor, Electronics, College of Engineering
Pune (COEP), An Autonomous Institution of Government of Maharashtra
Pune, India

Mehul S. Raval, Ph.D.

Associate Professor, Institute of Engineering and Technology
Ahmedabad University
Ahmedabad, Gujarat, India

Yogesh H. Dandawate, Ph.D.

Professor, Department of Electonics and Telecommunication Engineering
Vishwakarma Institute of Information Technology
Pune, India

Kalyani R. Joshi, Principal, Ph.D.

Professor in Electronics and Telecommunication Engineering
P.E.S. Modern College of Engineering
Pune, India

Shilpa P. Metkar, Ph.D.

Assistant Professor, Electronics, College of Engineering
Pune (COEP), An Autonomous Institution of Government of Maharashtra
Pune, India

CRC Press
Taylor & Francis Group
Boca Raton London New York

CRC Press is an imprint of the
Taylor & Francis Group, an **informa** business

A CHAPMAN & HALL BOOK

CRC Press
Taylor & Francis Group
6000 Broken Sound Parkway NW, Suite 300
Boca Raton, FL 33487-2742

First issued in paperback 2019

© 2015 by Taylor & Francis Group, LLC
CRC Press is an imprint of Taylor & Francis Group, an Informa business

No claim to original U.S. Government works

ISBN-13: 978-1-4822-2822-9 (hbk)
ISBN-13: 978-0-367-37816-5 (pbk)

Visit the Taylor & Francis Web site at
http://www.taylorandfrancis.com

and the CRC Press Web site at
http://www.crcpress.com

Contents

Preface

This book is intended primarily for courses on image compression techniques for undergraduate through postgraduate students, research scholars, and engineers working in the field. It presents the basic concepts and technologies in a student-friendly manner. The major techniques in image compression are explained with informative illustrations, and the concepts are developed from the basics. Practical implementation is demonstrated with MATLAB® programs.

While teaching and contributing to research in the field, the authors felt the need for a book that provides a strong foundation and presents practical applications. Encouraged by our teaching and research experience and our interaction with the industry, we decided to embark upon this venture to provide students and researchers with a book that not only enables learning from the basic principles but can also be applied in real practical applications.

The Internet has become an effective and important medium in applications including entertainment, business, research, and education. The amount of data generated is in the thousands of trillions of bytes. For efficient storage of data on devices, mediums and transmission over a network, the data need to be compressed. Presently, significant costs are also involved with data storage. For these reasons, we have always been motivated toward the development of techniques for data compression. The authors have worked in the field for many years and have contributed to research in the field. When contacted to take up the project of writing a book on image compression techniques as a joint effort, the authors agreed to write chapters pertaining to their areas of expertise. This book covers all of the techniques used for image and video compression in a lucid manner along with MATLAB programs that students can use and understand practically. In order to evaluate the performance of the compression algorithms, quality analysis is performed. The quality metrics used are presented in a separate chapter. We have also introduced a chapter on compressed sensing, which is a modern technique for retaining the most important part of the signal during its acquisition. The chapter coverage is as follows.

Chapter 1 introduces the need to compress images and videos; the lossless and lossy taxonomies; along with advantages, disadvantages, and application areas of compression.

Chapter 2 deals with lossless compression techniques that retain the entire image information while compressing it. Even though these techniques have less compression ratios, they are still used in the last stage of lossy compression techniques, as well as in hybrid techniques. They are also used in the compression of medical images. The techniques of Huffman, arithmetic, and dictionary coding are discussed in detail and relevant examples are provided.

Frequency domain is the most common representation for image storage and display. It is useful to visualize certain feature(s) of a signal in a transform domain when they are not visible in the parent domain. Chapter 3 deals with image transforms and covers the fundamentals of image transforms. It discusses the discrete cosine transform (DCT), discrete Fourier transform (DFT), Walsh-Hadamard transform (WHT), and an optimal Karhunen-Loeve transform (KLT). It also provides the application for an image transform in the domain of digital watermarking.

Discrete cosine transform–based image compression generates a blockiness artifact at high compression ratios. Image compression using wavelet transforms overcomes this drawback. Multiresolution analysis (MRA) is possible with scaling and wavelet functions. An image is decomposed into low- and high-frequency coefficients at multiple levels, and threshold is applied on them to achieve the desired compression. Wavelet transform theory and its application toward image compression, and JPEG 2000, today's popular compression standard, are presented in Chapter 4.

Chapter 5 deals with image compression using vector quantization. It is based on Shannon's rate distortion theory, which states that better compression is achieved if samples are coded using vectors instead of scalars. The finite vectors of pixels are stored in memory called *codebook*, which is used during the coding and decoding of the images. The image to be compressed is divided into blocks called *input vectors* that are compared with vectors in memory called *codevectors* for matches based on some distance criteria. If a codevector matches an input vector, an index or address of memory location is stored or transmitted. An address has fewer bits, so compression is achieved. Decoding is the opposite of encoding. Algorithms, such as Linde-Buzo-Gray (LBG) and its variants, mean-removed, gain-shape, and self-organizing feature maps are explained in detail. The programs for designing and using vector quantizers are also presented along with quality analysis.

Chapter 6 briefly presents the basic concepts of digital video coding. The chapter also focuses on the development of different video coding standards from H.261 to the recent HEVC standard.

Due to the compression artifacts, the quality of images degrades. The quality also suffers during acquisition, transmission, and reproduction and processing. So, the measurement of image quality is important in numerous image-processing applications. The extent to which quality is degraded should be analyzed for further improvement in the performance of the compression scheme. The full-reference quality metrics such as traditional mean square error, peak signal-to-noise ratio, image fidelity, structural similarity, and others, are explained in Chapter 7. During compression, if a reference image is not available, then the technique known as *reduced reference image quality assessment* is used.

Chapter 8 deals with the newer domain in compression, known as *compressed* or *compressive sensing* (CS), which is a mathematical theory of measuring and retaining the most important part of the signal while it is being sensed. CS effectively performs dimensionality reduction of a signal in a linear manner, which results in data compression. The chapter focuses on the finite dimensional sparse signals and provides an overview of the basic theory underlying the ideas of compressive sensing. In the latter part of the chapter, an application based on compressive sensing theory is covered. This chapter is aimed at both the theorist and the practitioner. It will be a review for the novice, who is interested in peeking through the domain; and also act as a quick reference to the theorist.

The authors feel that after referring to this book, students and researchers will definitely feel comfortable and be motivated to further explore the exciting domain of image and video compression. Recent and past research papers from reputed journals were referred to while preparing the content for the book chapters. These are listed as comprehensive bibliographic material at the end each chapter. We look forward to feedback from the readers for further improvement of the book content.

MATLAB® is a registered trademark of The MathWorks, Inc. For product information, please contact:

The MathWorks, Inc.
3 Apple Hill Drive
Natick, MA 01760-2098 USA
Tel: 508 647 7000
Fax: 508-647-7001
E-mail: info@mathworks.com
Web: www.mathworks.com

Authors

Madhuri A. Joshi, Ph.D., is professor of electronics at the College of Engineering, Pune. She has more than 38 years of teaching experience. Dr. Joshi has been a commonwealth scholar for research at the Manchester Institute of Science and Technology, Manchester; and the recipient of the following: Best Teacher Award 2010–2011 from the Maharashtra Government, India; IETE–Professor S.V.C. Aiya Memorial Gold Medal (2011) for excellence in research; and the Award for Excellence in Telecom Education (2011). She has authored three books, *Electronic Components and Materials* (Shroff Publishers, 2004), *Digital Image Processing, an Algorithmic Approach* (PHI Learning Pvt. Ltd., 2006), and *Introduction to Embedded System Design* (Indian Society for Technical Education, 2010). In addition, she has written 137 papers including 75 at the international level.

Mehul S. Raval, Ph.D. (electronics and telecommunication engineering), is an associate professor at the Institute of Engineering and Technology (IET), Ahmedabad University, with a research interest in image and signal processing. He has more than 17 years of teaching and research experience. Dr. Raval is the recipient of the Best Engineering College Teacher Award for the State of Gujarat from the Indian Society for Technical Education and a Young Scientist Fellowship from the Government of Gujarat in 2005. He is a senior member of IEEE and has published extensively in peer-reviewed journals and conferences. Currently, Dr. Raval serves as the vice chair for the IEEE Gujarat Section.

Yogesh H. Dandawate, Ph.D. (electronics and telecommunication engineering), received a bachelor of engineering from the University of Pune (India) in 1991, a masters of engineering from Gulbarga University (India) in 1998, and a Ph.D. in electronics and telecommunications engineering from the University of Pune (India) in 2009. Presently, he is a professor in the Department of Electronics and Telecommunication Engineering at Vishwakarma Institute of Information Technology, Pune. Dr. Dandawate has 22 years of teaching experience and has published more than 43 papers in reputed national and international conferences and referred journals. His areas of interest include signal and image processing, embedded systems, and soft computing. He is senior member of the IEEE, and a fellow member of IETE, India.

Kalyani R. Joshi, Ph.D., is principal and professor of electronics and telecommunications at P.E.S. Modern College of Engineering, Pune. She has more than 24 years of teaching experience. Dr. Joshi has authored 27 papers, including 11 at the international level.

Shilpa Prabhakar Metkar, Ph.D., is an assistant professor with the Electronics and Telecommunications Engineering Department, College of Engineering, Pune (an autonomous institute of the Government of Maharashta). She is the recipient of the gold medal in third year, and third university ranker in final year of engineering. Dr. Metkar has 10 years of teaching experience. She is a recipient of the prestigious Career Award for Young Teachers (2013) and the All India Council of Technical Education and the Institution of Engineers—Young Engineers Award (2013–2014). Dr. Metkar has published six papers in national conferences, fourteen papers in international conferences, and four papers in international journals.

1

Introduction to Image Compression

Image compression plays a significant role in the storage and transmission of image data. In addition to image data, audio data, voice data, animations, full motion video, and graphics are also sent. This is called *multimedia data*. Image and video data occupy a large portion of bandwidth during transmission. Thus, image data need to be compressed without or with small loss of quality. The major challenge is to develop efficient image compression algorithms.

1.1 Need for Coding

An image may contain information that is redundant and or irrelevant. Removal of this data provides image compression. An image may contain redundant information and neighboring pixels may be correlated. Redundancy and irrelevancy reduction are the two fundamental components of compression. A part of the data is not relevant to the human visual system. Removal of this data is called *redundancy reduction*. Reduction of the data that is not relevant to the human visual system (HVS) is called *irrelevancy reduction*. The redundancy of images can be found from their statistical properties. The irrelevancy of the observer of images is related to the HVS. In image compression, redundancies are classified into three types: coding, interpixel, and psychovisual. Coding redundancy occurs when code words are not optimal. Interpixel redundancy is due to correlations between the pixels. The HVS rejects information that is visually not necessary, resulting in psychovisual redundancy. The reconstructed image can be obtained from the compressed data. This process is called *decompression*. Removal of the spectral and spatial redundancies leads to a reduction in the number of bits. It thus provides compression. However, compression ratio and the quality of a decompressed image are the trade-off factors.

1.2 Measurement of Quality

In lossy compression, the reconstructed image and the original image are almost the same. The measurement of quality of an image is a vital issue, particularly in video and image compression. Fundamentally, there are two

FIGURE 1.1
Transmitter.

approaches to estimate quality: reference-based quality measures and no-reference-based quality measures. In the first measure, an input reference image is available for comparison, and in the second, a directly reconstructed image is available at the output.

An image coder system contains three components: a source encoder, quantizer, and entropy encoder. Image compression consists of using transform techniques to de-correlate the source such as an image or video. The transform coefficients are quantized. The next step is to convert the quantized values to entropy. This is shown in Figure 1.1. Compression techniques are divided into two major categories: loss-/lossless-based compression and predication-/transform-based compression.

1.3 Lossless Compression

Lossless compression provides an image that is identical to the original image. Huffman coding, arithmetic decomposition, Lempel-Ziv, and run-length encoding (RLE) are some of the lossless image compression techniques.

In high-performance applications such as medical imaging and nondestructive testing, it is essential to recover the original images after lossless image decompression.

Lossless compression consists of two stages: de-correlation and entropy coding. The de-correlation step removes spatial redundancy by using techniques like run-length coding and transform techniques. Entropy means information. This step removes coding redundancy by using techniques like Huffman coding, arithmetic coding, and Lempel-Ziv-Welch (LZW) coding. There are image compression standards such as JPEG-LS, which provide a balance between complexity and efficiency. Effective compression is essential for low as well as high compression ratios.

1.4 Lossy Compression

Algorithms that restore the presentation to be similar but not the same as the original image are called *lossy techniques*. Such reconstruction of an image is an approximation of the original image. There is a need for the measurement of the quality of the image for the lossy compression technique. It provides a high compression ratio and is applied in video and image compression more often than lossless techniques. It is classified into frequency-oriented techniques, prediction-based techniques, and importance-oriented techniques. The image is transformed to a spatial domain or a frequency domain such as a discrete cosine transform (DCT), discrete Fourier transform (DFT) subband decomposition, or wavelet transform, which provides greater data compression at the expense of greater computation, and is called a *frequency-oriented technique*. The basis for compression uses other characteristics of images, such as filtering and lookup tables, subsampling, bit allocation, and quantization such as scalar and vector, which are called *importance-oriented techniques*.

The channel capacity depends on the lowest rate at which reliable communication can be established. Appropriate encoding and decoding systems can be used to achieve the desired channel capacity in practice. The limits to which data can be compressed are provided by using Shannon's source coding theorem. Data compression is not possible when the code rate is less than the Shannon entropy of the source. It may be possible at such a rate that the information will be lost. The rate distortion function is shown in Figure 1.2. It is, however, possible to get the code rate approximately close to the Shannon entropy, with very less probability of loss.

Entropy is a measure of information. It provides an absolute limit on the best possible lossless compression of any communication.

Lossy data compression theory is also developed by Shannon.

Lossless image compression techniques are discussed in Chapter 2.

The basic purpose of transform coding is to achieve de-correlation and energy compaction. Due to this property, the number of bits to be allocated to an image is less and thus image compression is possible. This is discussed in Chapter 3.

The wavelet transform has the flexibility of scaling in the frequency and time domain, simultaneously. This is covered in Chapter 4.

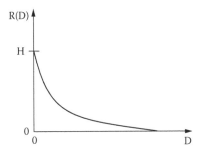

FIGURE 1.2
Rate distortion function.

Vector quantization (VQ) is based on Shannon's rate distortion theory, which says that better compression is achieved if samples are coded using vectors instead of scalars, which is explained in Chapter 5.

Video compression deals with reducing the number of bits required for processing a video sequence. While doing so, good video quality is maintained. Video compression is discussed in Chapter 6.

Digital images suffer due to a large number of distortions while processing. The resulting quality may not be good. It is thus important to measure quality in image-processing applications. This is discussed in Chapter 7.

Compressive sensing is a mathematical theory of measuring and retaining the most important part of a signal while sensing it. It effectively performs dimension reduction of the signal in a linear manner.

Bibliography

C. S. Burrus and R. A. Gopinath, *Introduction to Wavelets and Wavelet Transform*, Prentice Hall, Upper Saddle River, NJ, 1998.

T. M. Cover and J. A. Thomas, *Elements of Information Theory*, Wiley, Hoboken, NJ, 2006.

Y. H. Dandawate and M. A. Joshi, Color image compression using enhanced vector quantizer designed with self organizing feature maps, *International Conference in Information Processing (ICIP-07)*, Bangalore, India, August 2007.

M. Joshi, *Digital Image Processing: An Algorithmic Approach*, PHI Learning, Delhi, India, 2007.

M. A. Joshi and K. S. Jog, Analysis of subband vector quantisers for colour images, *IEEE TENCON 98*, New Delhi, 1998.

M. A. Joshi and K. R. Joshi, Objective quality measure in compression systems, *SPCCN1.1*, National Conference, VIT, July 1–3, 2005.

M. A. Joshi and M. B. Khambete, Priority based VQ using wavelet domain, *TENCON 2003*, Bangalore, India, 2003.

M. Marimuthu, R. Muthaiah, and P. Swaminathan, Review article: An overview of image compression techniques, *Research Journal of Applied Sciences, Engineering and Technology* 4(24): 5381–5386, 2012.

M. S. Nagmode and M. A. Joshi, Moving object detection using wavelet transform and dispersion centroid tracking in video sequences, *SPCCN01-2005*, V.I.T, Pune, pp. 81–85, July 1–4, 2005.

M. Nelson and J. -L. Gailly, *The Data Compression Book*, M&T Books, New York, 1996.

A. Papoulis and S. Unnikrishna Pillai, *Probability, Random Variables, and Stochastic Processes*, McGraw-Hill, New York, 2002.

D. Salomon, *Data Compression*, Springer, New York, 2003.

2

Lossless Image Compression

2.1 Introduction

Lossless image compression finds applications in medical imaging, industrial applications such as nondestructive testing, satellite imaging and remote sensing, real-time applications, document management systems, image archiving, HDTV, and in applications where image quality after decoding and reconstruction is very good, or no loss of information is expected.

Today, medical images are not stored on film, making image compression pivotal. Medical applications use lossless compression instead of lossy techniques, largely because of legal reasons, even though they provide lower compression rates compared to lossy compression.

Nondestructive testing (NDT) is a vital part of the quality control process in the manufacturing industries to detect defects in a manufactured component without physically destructing it. Using lossless compression techniques, such challenges can be addressed.

The remotely sensed satellite images have large data volume. This needs to be significantly reduced before transmission to earth. The quality of such images is needed to be as high as possible, because the images are to be used for further scientific analysis. Thus, a technique for image compression that is lossless is required.

Large numbers of applications involve huge data storage and transmission. In India, we have launched several satellites to collect and send information to the ground station in connection with the forecasting of weather conditions. These satellites are functioning round the clock, transmitting data to the ground station periodically. The received data are stored throughout the day and all days in the year. The data stored for a period of time are analyzed and used for forecasting the weather. Thus, we need large storage media to store the data received from the satellite. In practice, employing large size memory devices involves high cost. Hence, compressing and storing the data has become a necessity.

Document image compression is a research area that deals with the compression of images of scanned color documents. These documents generally have a very high resolution and thus are high quality. Such compressed documents can be transmitted at a high speed even over low-speed connections.

It is also possible to have authentic reproduction of the document with respect to fonts, color, paper texture, and pictures. While compressing, it is necessary to separate the text and figures from the images. The drawing may require high spatial resolution and background of a low resolution. So, both can be coded with a different rate of coding.

A typical lossless compression system consists of similar components as for lossy technique (i.e., A/D convertor, signal decomposition, quantization, and lossless coding). A/D converter samples and finely quantizes an image, producing digital representation in terms of x and y coordinates. Signal decomposition uses linear transforms. Such decompositions serve to compact the energy into a few coefficients.

The motivation to develop new techniques and stimulation for the rapid growth of research efforts contribute to the large commercial potential of image and video coding. It is difficult to achieve a large compression ratio along with good quality of an image after reconstruction. This also increases the complexity of the receiver. The characteristics of an ideal image coder are high fidelity in the image reconstructed at the receiver, low bit rate, and reduced complexity of the encoder and the decoder. Input redundancies can be removed during encoding. The encoded data are transmitted. Decoding is done at the receiver.

2.2 Source Encoders and Decoders

There is redundancy in an image due to psychovisual or interpixel aspects and coding. We can use a source encoder to reduce it. The source encoder should be designed in such a way that it is capable of eliminating coding, psychovisual, and interpixel redundancies in the input image. In the source encoder, the first block is called the *mapper,* which converts the given input image into a form that reduces interpixel redundancies. It is generally a reversible process. The next block is a *quantizer block,* where the psychovisual redundancies of the image are minimized. The third block is the *symbol coder* that creates a code to assign the quantizer output. The most frequently occurring values are assigned the shortest code words. This reduces the coding redundancy. Removing the different types of redundancies existing in the input data is the main function of the source encoder.

The source decoder, shown in Figure 2.1, contains two blocks. The quantizer block is not present because it is irreversible. The other two blocks have an inverse operation compared to the encoder.

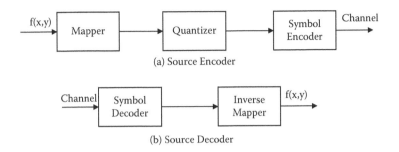

FIGURE 2.1
(a) Source encoder and (b) source decoder.

2.3 Coding Redundancy

The gray levels in any image are random in value. An image can be represented in the form of a histogram of the gray levels, which specifies the number of pixels in each gray level. Each gray level is represented using a constant number of bits, for example, eight bits. An image of size $a \times b$ requires $a \times b \times 8$ bits of memory to store the image. It is possible to calculate the number of bits necessary to represent each pixel using the following equation:

$$B_{avg} = \sum_{K=0}^{L-1} b(g_k) \cdot p(g_k) \tag{2.1}$$

where $b(g_K)$ represents the number of bits necessary to show Kth gray level, and $p(g_K)$ represents the probability associated with the Kth gray level.

When we represent the image using natural m-bit binary code then $b(g_K)$ = m bits then b_{avg} = m bits because the sum of $p(g_K)$ = 1. If we represent the gray levels of the image using a different number of bits for different gray levels based on the probability associated with the gray levels, as a result the average word length required to represent the image is reduced. As an example, if we use less bits to represent those gray levels whose probability is high and more bits for gray levels whose probability is less, this will yield less average number of bits compared to natural binary code. The data in Table 2.1 are used to show this fact.

The table contains two different codes, namely Code 1 and Code 2. Code 1 uses natural binary representation for the pixel gray levels, whereas Code 2 employs variable length coding.

TABLE 2.1

Fixed and Variable Length Code Example

g_k	$p(g_k)$	Code 1	$b_1(K)$	Code 2	$b_2(K)$
$g_0 = 0$	0.12	000	3	11	2
$g_1 = 1/7$	0.22	001	3	01	2
$g_2 = 2/7$	0.18	010	3	10	2
$g_3 = 3/7$	0.16	011	3	001	3
$g_4 = 4/7$	0.1	011	3	0001	4
$g_5 = 5/7$	0.2	101	3	00001	5
$g_6 = 6/7$	0.01	110	3	000001	6
$g_7 = 7/7$	0.01	111	3	00000001	7

2.4 Interpixel Redundancy

In this section, another important form of data redundancy called *interpixel redundancy*, which directly relates to the correlations between pixels in an image, is discussed. This is explained with an illustrative example.

Consider the image shown in Figure 2.2a.

The blocks in the image are scattered randomly. Another image shown in Figure 2.2b contains blocks arranged in two rows at regular intervals. The histograms for these two images are shown in Figure 2.2c,d which are identical for the two images. Employing a variable length coding process for these two images will not alter the level of correlation between pixels in these two images. Figure 2.2e,f shows the autocorrelation coefficients computed along one line of each image.

2.5 Psychovisual Redundancy

Luminance of a source is perceived by the human eye as brightness mainly because of reflection from the perceived object (e.g., human eye neglects intensity variations while viewing an object). Sometimes information such as the shape of an object is given more importance than its detailed texture. The human eye is selective in processing visual information. Information perception of this type is said to have psychovisual redundant data.

The examples of psychovisual redundancy are shown in Figure 2.3.

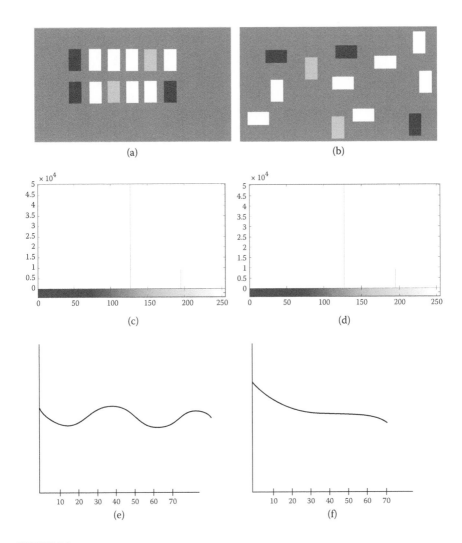

FIGURE 2.2
Interpixel redundancy. (a) Blocks arranged in a zigzag way. (b) Blocks arranged in two horizontal rows. (c) Histogram of (a). (d) Histogram of (b). (e) Autocorrelation coefficients of (a). (f) Autocorrelation coefficients of (b).

2.6 Image Compression Models

The general compression model consisting of two major building blocks is shown in Figure 2.4. The two major building blocks are encoder and decoder.

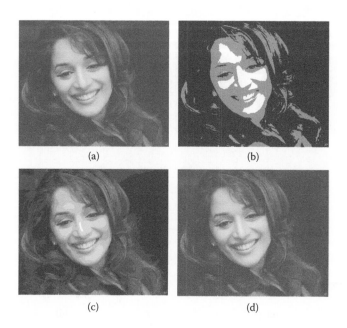

FIGURE 2.3
Psychovisual redundancy. (a) Image with 256 gray levels. (b) Image quantized using two levels. (c) Image quantized using eight levels. (d) Image quantized using 50 levels.

The input image $f(x,y)$ is fed to the encoder, and the encoder creates a set of symbols for the input data. These symbols are then transmitted through the channel and provided as input to the decoder. The decoder in turn decodes the data received and reconstructs an image $f'(x,y)$. The reconstructed image $f'(x,y)$ may be the same or may not be compared to $f(x,y)$. If $f'(x,y)$ is an exact replica of $f(x,y)$, then there is no distortion in the reconstructed image. Otherwise some level of distortion will be present in it. The encoder consists of two subblocks: source encoder and channel encoder. The source encoder is responsible for removing the data redundancy, whereas the channel encoder helps in error-free transmission. If the channel is noise-free, then there is no need to use a channel encoder. Similarly, the decoder consists of two subblocks: channel decoder and source decoder. The channel decoder detects the errors and corrects them. For a noise-free channel, the channel decoder

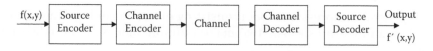

FIGURE 2.4
The general compression model.

is not necessary. The source decoder receives the error-free compressed data and reconstructs the image $f'(x,y)$, which is similar to the original image.

2.7 Channel Encoder and Decoder Realization

The role of the channel encoder and decoder is to reduce the impact of channel noise when the channel is noisy. The output of the source encoder contains less redundancy. It is highly sensitive to noise. So the channel encoder introduces a controlled form of redundancy into source encoder data to reduce the impact of noise. The important channel encoding technique that is used is called the *Hamming code*. This is based on appending enough bits to the data being encoded to ensure that in case of any error only a minimum number of bits will need to be changed to restore valid code words.

For example, the scientist, Richard Hamming, showed that if three bits of redundancy are added to a 4-bit word, the distance between any two valid code words is three, and all single bit errors could be detected and corrected. The 7-bit Hamming code word is denoted as $h_1, h_2, ..., h_7$. Out of these seven bits, four bits are used for information and they are denoted as $b_3 b_2 b_1 b_0$.

2.8 Information Theory

2.8.1 Information

Information theory provides some important concepts that are useful in digital image compression and representation of images, image transforms, and image quantization. Let us consider an image that has discrete gray levels and is denoted as r_k, where k can take values from 0 to L, where L is the number of pixels in the image (e.g., if the image is 16×16 then k takes values from 0 to 255). The probability associated with each gray level can be denoted as P_k, $k = 0, 1, 2, ..., L-1$. Then the information associated with r_k is given as in Equation (2.2):

$$I_k = -\log_2 P_k = 1/\log_2 P_k \tag{2.2}$$

Since

$$\sum_{K=0}^{L-1} P_k = 1 \tag{2.3}$$

each $P_k = 1$, and I_k is positive. From this, one can understand that the information conveyed is large when an unlikely message is generated.

2.9 Classification

Lossy and lossless image compression are two major categories of image compression:

- Lossless image compression preserves the precise data content of the input image.
- Lossy image compression may preserve the actual data to such an extent so as to get a specific image quality.

2.9.1 Compression Ratio

The level of image compression that can be achieved is represented by the *compression ratio*. The compression ratio is equal to the size of the original image compared to the size of the compressed image. A higher value ratio indicates better compression. In most cases, it is necessary to maximize the compression ratios while still meeting the quantities such as time to compress, time to decompress, computational cost, and quality. Image compression and decompression operations are said to be symmetrical operations if the time required to compress and decompress is the same. When one operation takes a longer time than the other, then it is called *asymmetrical compression*.

2.9.2 Contents of a Picture

If the contents of a picture are known prior to starting image compression, choosing a suitable procedure is possible. The type of contents has underlying structure, thus making it possible to design different image compression algorithms. For example, text, pictures, and videos are specific types of images.

Text image—A text image contains letters with their frequency, which is varying. We can code more frequently occurring letters with shorter code length.

Picture image—An image consists of parts that have less complexity or detail and can be represented by a smaller number of bits. The more detailed contents in the image, however, require more bits to encode.

Video—A video consists of sequential frames of pictures, as we move from one frame to another only a small part of the image or picture changes. Encoding the change is necessary.

Lossless compression methods include the following:

1. Shannon-Fano coding
2. Huffman coding
3. Lempel-Ziv coding
4. Arithmetic coding
5. Run-length coding

2.9.2.1 Shannon-Fano Coding

Shannon-Fano coding is a method that generates a binary tree. The probability of each symbol is found, and a code is assigned with a corresponding code length. Even though Shannon-Fano is not an optimum coding method, it is a computationally less complex algorithm.

An example of Shannon-Fano coding is provided in Table 2.2.

The steps are as follows:

1. First prepare an ordered table giving the frequency of each symbol.
2. Then divide each part of the table into two segments.
3. Ensure while dividing that both the parts have nearly the same sum of frequency.
4. Now repeat the procedure until a single symbol is left.

Algorithm for Shannon-Fano coding:

1. First prepare an ordered table giving the frequency of each symbol.
2. Symbols are begun as per frequency in descending order.
3. Then divide the segment into two parts, both having nearly the same sum of frequencies.
4. Next assign a binary code 0 to the code word of the upper part and 1 to the lower part.
5. Last, search for the next segment containing more than two symbols and repeat division.

TABLE 2.2

Shannon-Fano Coding

Symbol	Frequency	Code Length	Code	Total Length
X	12	2	00	24
Y	6	2	01	12
Z	5	2	10	10
T	4	3	110	12
E	4	3	111	12

Notes: Total symbols = 31. Shannon-Fano coding: 70 bits. Linear coding: 92 bits.

Figure 2.5 and Table 2.3 provide examples of implementation of an algorithm of Shannon-Fano coding.

2.9.2.2 Huffman Coding

The simplest approach used to compress the image data without any loss is variable length coding as it removes the coding redundancy. For constructing a variable length code, the probabilities associated with different gray

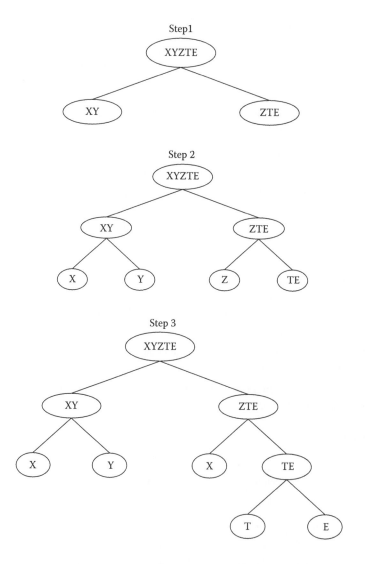

FIGURE 2.5
Shannon-Fano coding.

TABLE 2.3

Shannon-Fano Coding: Steps in the Construction of Code

Symbol	Frequency	Step 1		Step 2		Step 3	
		Sum	Code	Sum	Code	Sum	Code
X	12	12	0	12	00		
Y	6	18	0	6	01		
Z	5	13	1	5	10		
T	4	8	1	8	11	8	110
E	4	4	1	4	11	4	111

levels are required. The Huffman coding technique provides optimum code symbols per source symbol. This is illustrated in Table 2.4.

Huffman code is used in image formats (e.g., ZIP, JPEG). Huffman coding is designed such that the probability of occurrence of every symbol results in its code length. Then a binary tree is generated.

Construction of the tree is as follows:

1. Search for two nodes having the lowest frequency which are not yet assigned to a parent node.
2. Couple these two nodes together to a new interior node.
3. Add both frequencies and assign this value to the two interior nodes.
4. Repeat until all nodes are combined together in a root node.

Steps in the generation of the tree are provided in Figure 2.6.
Table 2.5 provides an example of coding and decoding of a symbol as follows:

Coding:

chinchinchua

0 10 110 1111 0 10 110 1111 0 10 11100 11101

TABLE 2.4

Huffman Coding Example

Symbol	Frequency	Code
c	3	0
h	3	10
i	2	110
n	2	1111
u	1	11100
a	1	11101

Notes: As per our steps, *u* and *a* are coupled in the first step. The new interior node will get a frequency of 2.

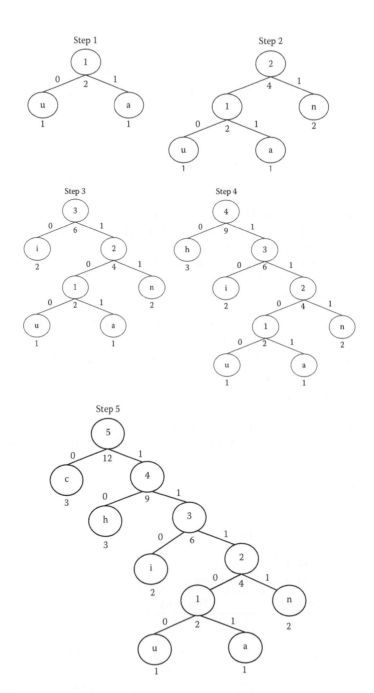

FIGURE 2.6
Huffman coding—tree generation.

TABLE 2.5

Huffman Coding–Decoding Example

Encoded	Decoded
0	c
10	h
110	i
1111	n
0	c
10	h
110	i
1111	n
0	c
10	h
11100	u
11101	a

Encoded data: 33 bits

Original data: 36 bits with 3 bits for each symbol

Decoding:

The Huffman tree is passed through with the encoded data step by step. When a node is reached, where it is not having a successor the assigned symbol will be written to the decoded data.

```
0 10 110 1111 0 10 110 1111 0 10 11100 11101
```

2.9.2.3 Lempel-Ziv 77

Lempel-Ziv 77 (LZ77) is an algorithm that derives a sequence out of the previous contents and not from the original image.

Coding is done as follows:

1. Address to already coded contents
2. Assign sequence length
3. Code denoting symbol

If no identical byte sequence is available from former contents, assign the address 0, and the sequence length 0, and code the new symbol.

Huffman code requires simple probabilities; hence, for real-time applications, Huffman encoding becomes impractical as source statistics are not always known *a priori*. A code that uses statistical interdependence of the letters in the alphabet in addition to their individual probabilities of occurrences might be more efficient and is called the *Lempel-Ziv algorithm*. This algorithm belongs to the class of universal source coding algorithms.

TABLE 2.6

Lempel-Ziv Coding–Decoding Example

Numeric position	1	2	3	4	5	6	7	8	9
Subsequence	0	1	00	01	011	10	010	100	101
Numerical representation			11	12	42	21	41	61	62
Binary coding sequence			0010	0011	1001	0100	1000	1100	1101

Lempel-Ziv universal coding (see Table 2.6) is based on the following principles:

1. The compression of an arbitrary binary sequence is possible by coding a series of zeros and ones as some previous such string (prefix string) plus one new bit.
2. The new string formed by such parsing becomes a potential prefix string for future strings.

These variables are called *phrases* (subsequences). The phrases are listed in a dictionary or codebook of the existing phrase in the codebook and append the new letter, which stores the existing phrases and their locations. In encoding a new phrase, specify the location of the existing phrase in the codebook and append the new letter.

Consider the following example: Determine the Lempel-Ziv code for the given sequence:

```
000101110010100101
```

The given sequence is

```
00, 01, 011, 10, 010, 100, 101
```

Depending on this, a table can be formed as

```
Maximum number in numerical representation = 6
6 = 110
No of bits = 3
```

that is,

```
0 = 000
1 = 001
2 = 010
4 = 100
5 = 101
6 = 110
```

The last symbol of each subsequence in the codebook is an innovation sequence corresponding to the last bit of each uniform block of bits in the binary-encoded representation of the data stream, representing the innovation symbol for the particular subsequence under consideration. The remaining bits provide the equal binary representation of the pointer in the root subsequence that matches the one in question except for the innovation number.

2.9.2.4 Arithmetic Coding

Arithmetic coding has the highest efficiency in coding symbols regarding the probability of occurrence. It provides a better compression rate compared to Huffman coding. It is more complex to implement. Arithmetic coding is used in JPEG format.

The problem of bit fractions is circumvented by arithmetic coding. For sending a word, its letters are assigned probabilities and are also assigned a unique range. The width of this range is equal to the probability of occurrence and is in the interval 0 to 1. The compression is achieved with narrowing the interval progressively with each new symbol. Initially, the interval width is 1; after the letter p is encoded the width reduces to 0.1. Adding letter o results in width reduced to 0.02 and this continues for the remaining letters added to the word.

Examples of arithmetic coding are presented in Tables 2.7 and 2.8.

2.9.2.5 Run-Length Coding

Run-length encoding is a lossless image compression technique. The gray level values, which are the same in a sequence, are represented as a single value and a count of its occurrence. This is most useful for data containing many such runs (e.g., images having continuous same gray level objects, drawings, etc.). The content of data restrict the compression rate. It is mainly used for encoding monochrome graphic data.

TABLE 2.7

Sample Symbol Ranges

Letter	Probability	Range
A	0.1	0.0–0.1
I	0.1	0.1–0.2
L	0.2	0.2–0.4
M	0.1	0.4–0.5
N	0.1	0.5–0.6
O	0.2	0.6–0.8
P	0.1	0.8–0.9
Y	0.1	0.9–1.0

TABLE 2.8

Scaled Symbol Ranges

Letter	Range	Interval	Interval after Encoded Symbol Width
P	0.8–0.9	1.0	0.0–1.0
O	0.6–0.8	0.1	0.8–0.9
L	0.2–0.4	0.02	0.864–0.868
Y	0.9–1.0	0.004	0.8676–0.8680
N	0.5–0.6	0.0004	0.86780–0.86784
O	0.6–0.8	0.00004	0.867824–0.867832
M	0.4–0.5	0.000008	0.8678272–0.8678280
I	0.1–0.2	0.0000008	0.86782728–0.867827360
A	0.0–0.1	0.00000008	—
L	0.2–0.4	0.000000008	—

Algorithm—Any sequence of identical symbols will be replaced by a counter identifying the number of repetitions and the particular symbol (e.g., sssss will be coded as 5s.

Example—For example, consider an image of a blackboard that has white characters on a black background. It will be composed of many long runs of black pixels in the blank space, and many short runs of white pixels within for the characters. Let us consider a single scanline, with B representing a black pixel and W representing white:

BBBBBBBBBBBBWBBBBBBBBBBBB WWWBBBBBBBBBBBBBBBBBBBBBBBBBBBBBWBBBBBBBBB
BBBBBB

By applying a run-length encoding algorithm to this line we get twelve Bs, one W, twelve Bs, three Ws, and so forth.

The result is that 67 characters in the original are represented in only 18 characters using the run-length code. The actual format used for the storage of images is generally binary rather than ASCII characters like this, but the principle remains the same. Even binary data files can be compressed with this method; file format specifications often dictate repeated bytes in files as padding space. However, newer compression methods such as DEFLATE often use LZ77-based algorithms, a generalization of run-length encoding that can take advantage of runs of strings of characters (such as BWWBWWBWWBWW).

Run-length encoding performs lossless image compression. One of the techniques used in Fax machines is run-length encoding.

Bibliography

M. V. Deshpande and M. Joshi, Digital controller, *International Conference of IEEE TENCON '94*, Singapore, August 1994.

M. Dipperstein, *Arithmetic Code Discussion and Implementation*, http://michael. dipperstein.com/arithmetic/index.html.

M. Joshi and K. S. Jog, A new approach for image compression using an image array, *IEEE TENCON '98*, New Delhi, India, pp. 630–633.

M. Joshi and K. S. Jog, Pattern recognition and image processing techniques, *Institution of Engineers (I), Computer Engineering Division, Seventh National Convention*, October 1991, pp. 121–134.

M. Nelson and J. -L. Gailly, *The Data Compression Book*, M&T Books, New York, 1996.

K. Sayood, *Introduction to Data Compression*, Elsevier, London, 2012.

S. Smith, The Scientists and Engineers Guide to Digital Signal Processing, Amazon. http://www.dspguid e.com/pdfbook.htm/.

3

Image Transforms

3.1 Introduction

According to the *Merriam-Webster* dictionary, *transform* in its broadest sense means "change in the external form or in the inner nature." Continuing, mathematical *transform* means: "to change to a different form having the same value." In this chapter, we are focusing on the image transforms. Usually, an image is represented as a function of two spatial variables (x, y) and represented as $f(x, y)$. The intensity at a particular point in an image is the value taken by the function $f(x, y)$ at that spatial location. This domain of representation is most common for image storage and display. Since an image is a representation in space through spatial coordinates, domain is termed as the *spatial* domain. The term *image transform* refers to the mathematical process of converting and representing an image into its alternative form. For example, an image can be represented as a series summation of sinusoids with varying degree of magnitude and frequencies by the cosine transform. This alternative representation of an image is known as *frequency domain*. A typical transformation process between spatial and frequency domain is shown in Figure 3.1.

From Figure 3.1, it should be noted that the conversion from the spatial to the frequency domain is known as a *forward* transform and the other way around is known as an *inverse* transform. As noted in the definition earlier, these transforms should be lossless, meaning that the information content of the signal should not be altered or lost due to the process of transformation.

An important question that must be pondered upon is: Why do we need these transforms when the information content should not change? These transforms help in visualizing the same information content but from a different perspective. It is useful to visualize certain feature(s) of a signal in transform domain when they are not visible in the parent domain. For example, information about the frequency content of the signal is available only in the frequency domain and not in its time domain visualization. In another reasoning, certain image processing operations can be best realized by transforming the signal, carrying out the desired operation, and returning to the parent domain. The general principle in applying the transform is depicted

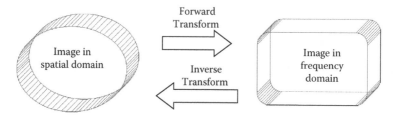

FIGURE 3.1
A transformation in between the spatial and frequency domain.

in Figure 3.2. Thus, input image is transformed from spatial domain to the frequency domain. In that domain the desired operation is carried out and then inverse transform is applied to go back to the spatial domain.

Image transform is useful in many applications and can be used for a range of purposes. Some of the important applications are as follows:

Image enhancement—Transform can be used to exploit limitations of the human visual system and improve perceptual quality and computational efficiency of the image.

Filtering—Transform can help in isolating certain components of interest in an image or can help to remove spurious noise components from the band of interest.

Image compression—Image transform helps in de-correlating the data, preserving the principal components of a signal, and quantizing other components to obtain the signal compression.

Pattern recognition—Image transform captures many features like edges, corners, and statistical moments to identify objects and/or to classify them.

Convolution—It is a computationally expensive operation in time or spatial domain, which in the frequency (transform) domain reduces to a multiplication operation by using an image transform.

FIGURE 3.2
The general principle in carrying out a specific operation in the transform domain.

3.2 Fundamentals of Image Transforms

In this chapter, we will consider the case of linear transforms, meaning that we get an output sequence (matrix) for the input sequence (matrix). Mathematically, transforms decompose the given two-dimensional (2D) signal (image) into a series summation of basis (*unitary*) matrices. Unitary matrices form the basis images in the transform. Unitary matrices encompass a very important class of matrices, and they have the following note-worthy properties like an identity matrix. They are as follows:

1. These matrices, when applied as a transformation over vectors, preserve their lengths.
2. They also preserve angle between the vectors.

Unitary matrices act as the identity transformations, and they form quite a large family in matrices. Before formally delving into the details of unitary matrices, let us recall some basic facts.

3.2.1 Orthogonal Functions

Definition 3.1: Consider a set of real valued continuous functions $\{f_n(t)\} = \{f_0(t), f_1(t), f_2(t)...\}$. This set is orthogonal over an interval $(t, t + T)$ if the following is fulfilled:

$$\int_{t}^{t+T} f_m(t) \, f_n(t) = \begin{cases} 0 & \text{if} \quad m \neq n \\ k & \text{if} \quad m = n \end{cases} \tag{3.1}$$

This $\{f_n(t)\}$ forms the set of real valued orthogonal basis functions. ∎

Definition 3.2: If $k = 1$ then the set is known as orthonormal—that is,

$$f_m(t)f_n^*(t) = \delta_{mn} = \begin{cases} 1 & m = n \\ 0 & m \neq n \end{cases}$$

where $f_n^*(t)$ indicates the complex conjugate of $f_n(t)$.

An orthogonal set of vectors can be made orthonormal by the following scaling:

$$f_n \rightarrow \frac{1}{(f_n f_n^*)^{\frac{1}{2}}} f_n$$

It can also be shown that every set of orthonormal vectors is linearly independent. ∎

3.2.2 Unitary Matrix

Definition 3.3: A matrix U is unitary if $U^{-1} = U^{*T}$, where U^* is a complex conjugate, and U^T is a transpose of U. Therefore, $U^{*T}U = UU^{*T} = I$, where I is the identity matrix. ∎

Some of the important properties of unitary matrices are as follows:

1. The columns of U are an orthonormal set (i.e., each column has a unit length).
2. The rows of U also form a unitary basis (orthonormal set); each row has a unit length.
3. The product of two unitary matrices is also a unitary matrix. All U^{*T}, U, I are unitary.
4. The inverse of a unitary matrix is also unitary.
5. The unitary matrix is diagonalizable:

$$U = \sum_{i=1}^{n} \lambda_i v_i v_i^T$$

where $\{\lambda_1, \lambda_2, \ldots, \lambda_n\}$ is a set of eigenvalues, and $\{v_1, v_2, v_3 \ldots v_n\}$ is the set of corresponding eigenvectors. This decomposition is known as the spectral decomposition of the unitary matrix.

3.2.3 Unitary Transform

Let us first consider the case of a one-dimensional (1D) signal for understanding. The signal can be decomposed as $x(t) = \sum_{n=0}^{\infty} C_n f_n(t)$, where C_n is known as the nth coefficient of expansion for an orthonormal set $C_n = \int_t^{t+T} x(t) f_n(t) dt$. Some of the important properties of $\{f_n(t)\}$ are as follows:

Definition 3.4: Completeness
   A set of an orthogonal function is complete or close if any one of the following holds:

1. A signal $x(t)$ has $\int x^2(t) dt < \infty$ (i.e., signal has a finite energy such that $\int x(t) f_n(t) dt = 0$ for $n = 0,1,2,\ldots$).

2. A signal $x(t)$ is piecewise continuous with finite energy (i.e., $\int x^2(t)dt < \infty$ then the minimum mean square error $\int |x(t) - \hat{x}(t)|^2 \, dt < \varepsilon$ converges). Then $\hat{x}(t) = \sum_{j=0}^{N-1} C_j f_j(t)$ is the approximate signal reconstruction using N terms, and the minimum mean square error tends to zero as $N \to \infty$.

Point 2 is more liberal for a set to be orthogonal than the condition in Point 1. This allows for a signal representation by a finite set of coefficients $\{C_0, C_1, C_2 \dots C_{N-1}\}$. ∎

3.2.4 One-Dimensional Signals

Considering the discrete samples of the signal $\mathbf{X} = \{x_1, x_2, x_3, \dots, x_N\}^T$, the transform can be written in the matrix as follows:

$$\mathbf{V}_{N\times 1} = \mathbf{U}_{N\times N}\mathbf{X}_{N\times 1} \tag{3.2}$$

where $\mathbf{V}$ is the transformed vector, $\mathbf{U}$ is the transformation matrix, and $\mathbf{X}$ is the signal vector.

The inverse transformation then can be written as

$$\mathbf{X}_{N\times 1} = \mathbf{U}_{N\times N}^{-1}\mathbf{V}_{N\times 1} = \mathbf{U}_{N\times N}^{*T}\mathbf{V}_{N\times 1} \tag{3.3}$$

3.2.5 Two-Dimensional Signals

Considering the case of an image $I(x,y); 0 \le x, y \le N - 1$, the forward transformation is represented as follows:

$$V(u,v) = \sum_{x,y=0 \, x,y=0}^{N-1 \quad N-1} U_{u,v}(x,y)I(x,y); 0 \le u, v \le N - 1 \tag{3.4}$$

where $V(u,v)$ is the forward transform of image $I(x,y)$, and $U_{u,v}$ is the unitary matrix for each (u,v) in V. This is known as the *forward transformation kernel*. Then (x,y) are the spatial variables with N rows and N columns of image I. Similarly, (u,v) are the transform variables. Here, N^2 is the number of unitary matrices required for image transformation.

Given the forward transform of an image $V(u,v)$, the inverse transformation of the image is represented in terms of unitary matrix as follows:

$$I(x,y) = \sum_{u,v=0 \, u,v=0}^{N-1 \quad N-1} U_{u,v}^*(x,y)V(u,v); 0 \le x, y \le N - 1 \tag{3.5}$$

where $U_{u,v}^*(x,y)$ represents the *inverse transformation kernel*. Thus, we can represent an image as a series combination of the unitary matrices. Equations (3.4) and (3.5) are called a *transform pair*. Like the 1D signals, if the image is reconstructed using the finite coefficients then the reconstruction error is as follows:

$$\epsilon^2 = \sum_{x,y=0}^{N-1} \sum_{x,y=0}^{N-1} \{I(x,y) - \hat{I}(x,y)\}^2 \tag{3.6}$$

This error can be minimized if $\{U_{u,v}(x,y)\}$ is complete, and it represents the set of orthonormal basis functions. As seen earlier, one of the important properties of orthonormal transform is that the inverse of the transform matrix is simply its conjugate transpose. Orthonormal transforms are energy-preserving ones; energy in the native domain is preserved in the transformed domain. This can easily be visualized in the case of 1D signal transformation for a real number.

$$\sum_{i=0}^{N-1} v_i^2 = v^T v \tag{3.7}$$

$= (UX)^T (UX) = X^T U^T UX$ (invoking Equation 3.2).
Now we know that U is a unitary matrix, and $U^T U = I$:

$$X^T U^T UX = X^T X = \sum_{i=0}^{N-1} x_i^2$$

Therefore,

$$\sum_{i=0}^{N-1} v_i^2 = \sum_{i=0}^{N-1} x_i^2 \tag{3.8}$$

The computational complexity of the image is very high. As seen from Equation (3.4), for every value of (u,v), the number of complex additions and multiplications to be performed is $\mathcal{O}(N^2)$. With $0 \leq x,y \leq N - 1$ the total number of complex additions and multiplications is $\mathcal{O}(N^4)$, which is very high. This exorbitant computational burden can be reduced if the image transform is *separable*. The transform is said to be separable if,

$$U_{u,v}(x,y) = U_u(x) \cdot U_v(y) = U_1(u,x) \cdot U_2(v,y) \tag{3.9}$$

If $U_1(u,x) = U_2(v,y) = U$, then the kernel is symmetric and U_1, U_2 will be unitary matrices—that is, $UU^{*T} = U^{*T}U = I$. The forward transformation in the case of separable transforms will take the following form:

$$V(u,v) = \sum_{x,y=0}^{N-1} \sum_{x,y=0}^{N-1} U_1(u,x)I(x,y)U_2(v,y); 0 \le u,v \le N-1 \qquad (3.10)$$

and the inverse transformation will take the form

$$I(x,y) = \sum_{u,v=0}^{N-1} \sum_{u,v=0}^{N-1} U_1^*(u,x)V(u,v)U_2^*(v,y); 0 \le x,y \le N-1 \qquad (3.11)$$

In this chapter, we are considering 2D transforms, which are separable. In that case we can take the 2D transform by first taking the transform along one of the dimensions (i.e., row/column) and then repeating the operation along the other (i.e., column/row). In matrix terminology, it is $V = UIU^T$ and $I = U^*VU^{*T}$. The computational complexity along one dimension is $\mathcal{O}(N^3)$ and thus along both the dimensions total computational complexity will be $\mathcal{O}(2N^3)$. This is significantly less than nonseparable transforms with $\mathcal{O}(N^4)$. Thus, it can be seen that the purpose of transformation is to represent an image in terms of the basis images. The important properties of unitary transforms are summarized as follows:

1. Energy-preserving property (i.e., $\|V\|^2 = \|X\|^2$), where $\|\cdot\|$ is the norm.
2. The unitary transform is a rotation of the signal vector into an N-dimensional space (i.e., basis coordinates are rotated and the angle between the vectors is preserved in the transform domain).
3. They also showcase the energy compaction property, and a large proportion of the energy is represented by a very small set of coefficients.
4. Interestingly, the transform coefficients are de-correlated.
5. The randomness (entropy) or the information content is preserved by the unitary transform.

3.3 Two-Dimensional Image Transforms with a Fixed Basis

In this chapter, we will deal with the linear transforms that use either the fixed orthonormal basis or derive the data from the basis. In the following sections, we will present a variety of transforms available to us and discuss the applications that pertain to them.

Often transforms evolve due to their coherence with specific application requirements (for example, in the application like data compression, transform effectiveness is determined by the amount of energy compaction provided by the transform). One of the important figures of merit to check the energy compaction of a particular transform is to calculate the ratio of the arithmetic mean of the variance of a transform coefficient to its geometric mean. This is known as the *transform coding gain*:

$$G_{TC} = \frac{\frac{1}{N}\sum_{i=0}^{N-1}\sigma_i^2}{\left(\prod_{i=0}^{N-1}\sigma_i^2\right)^{\frac{1}{N}}}$$

(3.12)

where σ_i^2 is the variance of the *i*th coefficient.

Transforms can be interpreted in several ways; however, in this chapter we are going to evaluate the transform as a decomposition of a signal into its basis set. One-dimensional signals can be expanded in terms of rows of the transform matrix. Two-dimensional signals can be seen as an expansion in terms of matrices formed due to the outer product of the rows of the transform matrix. The rows or columns of the transform matrix are called *basis vectors of the transform*, and they form the orthonormal set. As a first example of a 2D transform with a fixed basis, we will explore the Fourier transform.

3.4 Two-Dimensional Fourier Transforms

A 2D Fourier transform is a mathematical transformation to convert signals from the spatial domain to the frequency (Fourier) domain. The transform is reversible in a lossless fashion from the frequency to the spatial domain. Consider the case of a continuous image signal $f(x,y)$. The *forward Fourier transform* is given as

$$F(u,v) = \int\int_{-\infty}^{\infty} f(x,y)e^{-j2\pi(ux+vy)}\,dx\,dy$$

(3.13)

Similarly, the *inverse Fourier transform* is given as

$$f(x,y) = \int\int_{-\infty}^{\infty} F(u,v)e^{j2\pi(ux+vy)}\, du\, dv \tag{3.14}$$

It should be noted that throughout this chapter, we use (x,y) as spatial coordinates and (u,v) for frequency domain coordinates irrespective of the transforms. In case we are considering the 2D discrete Fourier transform (DFT) of an image $f(x,y)$ with size $N \times N$, the kernels are as follows:

$$F(u,v) = \frac{1}{N}\sum_{x,y=0}^{N-1}\sum_{x,y=0}^{N-1} f(x,y)e^{-j\frac{2\pi}{N}(ux+vy)} \tag{3.15}$$

where $0 \le u, v \le N - 1$.

$$f(x,y) = \frac{1}{N}\sum_{u,v=0}^{N-1}\sum_{u,v=0}^{N-1} F(u,v)e^{j\frac{2\pi}{N}(ux+vy)} \tag{3.16}$$

where $0 \le x, y \le N - 1$.

It can be seen that $e^{-j2\pi(ux+vy)}/e^{-j\frac{2\pi}{N}(ux+vy)}$ forms the forward transform kernel while $e^{j2\pi(ux+vy)}/e^{j\frac{2\pi}{N}(ux+vy)}$ forms the inverse kernel. Then $j = \sqrt{-1}$ makes this kernel complex. These kernels are also the basis image in this transform. The Fourier spectrum is computed as $|F(u,v)| = [R^2(u,v) + I^2(u,v)]^{\frac{1}{2}}$, where $R(u,v)$ and $I(u,v)$ are the real and imaginary parts of the complex coefficients. The power spectrum $P(u,v)$ is computed as

$$P(u,v) = |F(u,v)|^2 = [R^2(u,v) + I^2(u,v)] \tag{3.17}$$

We will now summarize the important properties of 2D DFT.

3.4.1 Separability

It is not difficult to show that Fourier kernels have the property of separability. It can also be observed that kernels have the property of symmetry as well. This means that 2D transforms can be computed using 1D transforms. Pictorially, this operation on $N \times N$ image $f(x,y)$ can be depicted as in Figure 3.3.

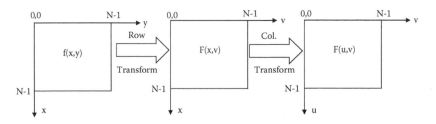

FIGURE 3.3
Application of a one-dimensional kernel to obtain the two-dimensional transformation.

It can be seen from Figure 3.3, that over an image $f(x,y)$, first the 1D kernel is applied row-wise to obtain the intermediate result $F(x,v)$. Then the same kernel is applied column-wise over the intermediate result to yield the final transformed image $F(u,v)$. The mathematically forward kernel can be realized as follows:

$$F(u,v) = \frac{1}{N} \sum_{x,y=0}^{N-1} \sum_{x,y=0}^{N-1} f(x,y) e^{-j\frac{2\pi}{N}(ux+vy)}$$

$$= \frac{1}{N} \sum_{x=0}^{N-1} e^{-j\frac{2\pi}{N}ux} \cdot N \cdot \frac{1}{N} \sum_{y=0}^{N-1} f(x,y) e^{-j\frac{2\pi}{N}vy}$$

Now,

$$\frac{1}{N} \sum_{y=0}^{N-1} f(x,y) e^{-j\frac{2\pi}{N}vy} = F(x,v) \text{ (Intermediate result)}$$

$$= \frac{1}{N} \sum_{x=0}^{N-1} F(x,v) e^{-j\frac{2\pi}{N}ux} \text{ (Final result)}$$

The same principle can also be applied to the inverse kernel due to the symmetry of the kernel:

$$f(x,y) = \frac{1}{N} \sum_{u,v=0}^{N-1} \sum_{u,v=0}^{N-1} F(u,v) e^{j\frac{2\pi}{N}(ux+vy)}$$

$$= \frac{1}{N} \sum_{u=0}^{N-1} e^{j\frac{2\pi}{N}ux} \cdot N \cdot \frac{1}{N} \sum_{v=0}^{N-1} F(u,v) e^{j\frac{2\pi}{N}vy}$$

Now,

$$\frac{1}{N} \sum_{v=0}^{N-1} F(u,v) e^{j\frac{2\pi}{N}vy} = f(u,y) \text{ (Intermediate result)}$$

$$= \frac{1}{N} \sum_{u=0}^{N-1} f(u,y) e^{-j\frac{2\pi}{N}ux} \text{ (Final result)}$$

3.4.2 Translation

A 2D image $f(x,y)$ is translated by (x_0,y_0) to form the translated version $f(x-x_0, y-y_0)$. Let the Fourier transform of this version be $F_{u,v}^{trans}$:

$$F_{u,v}^{trans} = \frac{1}{N} \sum_{x=0}^{N-1} \sum_{y=0}^{N-1} f(x-x_0, y-y_0)\, e^{-j\frac{2\pi}{N}[u(x-x_0)+v(y-y_0)]}$$

$$= \frac{1}{N} \sum_{x=0}^{N-1} \sum_{y=0}^{N-1} f(x-x_0, y-y_0)\, e^{-j\frac{2\pi}{N}(ux+vy)}\, e^{-j\frac{2\pi}{N}(ux_0+vy_0)}$$

$$F_{u,v}^{trans} = F(u,v) e^{-j\frac{2\pi}{N}(ux_0+vy_0)}$$

$$\therefore f(x-x_0, y-y_0) \rightarrow F(u,v) e^{-j\frac{2\pi}{N}(ux_0+vy_0)} \tag{3.18}$$

For the inverse Fourier transform,

$$F(u-u_0, v-v_0) \rightarrow f(x,y) e^{j\frac{2\pi}{N}(ux_0+vy_0)} \tag{3.19}$$

It can be seen from Equations (3.18) and (3.19) that translating the origin in spatial (frequency) by (x_0,y_0) leads to multiplication by the exponential term in the frequency (spatial) domain. It is interesting to note the relationship between the spatial and frequency domains.

3.4.3 Periodicity and Conjugate Properties

The DFT and IDFT are periodic with a period of N:

$$F(u,v) = F(u+N, v+N) \tag{3.20}$$

$$F(u+N, v+N) = \frac{1}{N} \sum_{x,y=0}^{N-1} \sum_{x,y=0}^{N-1} f(x,y) e^{-j\frac{2\pi}{N}(ux+Nx+vy+Ny)}$$

$$\times \frac{1}{N} \sum_{x,y=0}^{N-1} \sum_{x,y=0}^{N-1} f(x,y) e^{-j\frac{2\pi}{N}(ux+vy)} \cdot e^{-j2\pi(x+y)}$$

$$= \frac{1}{N} \sum_{x,y=0}^{N-1} \sum_{x,y=0}^{N-1} f(x,y) e^{-j\frac{2\pi}{N}(ux+vy)} = F(u,v)$$

Periodicity helps with visualizing the Fourier spectrum. For the case of a conjugate property, if $f(x,y)$ is a real valued function, then $F(u,v) = F^*(-u,-v)$ and $|F(u,v)| = |F(-u,-v)|$. A typical DFT spectrum is shown in Figure 3.4. The spectrum is plotted after placing $F(0,0)$ at the center of the spectrum. This is done because it is easy to visualize the spectrum symmetry around the center. It is also visible that the *dc* term $F(0,0)$ dominates the spectrum.

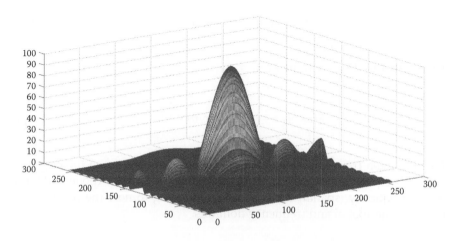

FIGURE 3.4
A typical Fourier spectrum of an image.

3.4.4 Rotation

Let us represent spatial coordinates *(x,y)* and frequency coordinates *(u,v)* in the polar coordinates:

$$x = r\cos\theta, \quad y = r\sin\theta$$

$$u = w\cos\varphi, \quad v = w\sin\varphi$$

Therefore, $f(x,y) \to f(r,\theta)$ and $F(u,v) \to F(w,\varphi)$, then $f(r,\theta+\theta_0) \leftrightarrow F(w,\theta+\theta_0)$. If an image *f(x,y)* is rotated by an angle θ_0, then transform $F(u,v)$ is rotated by the same angle. Similarly, rotation of the Fourier transform by an angle θ_0 results in the image being rotated in the spatial domain by the same angle. The above comments are validated when looking at Figure 3.5. Figure 3.5 illustrates the rotation property of the Fourier spectrum and an image.

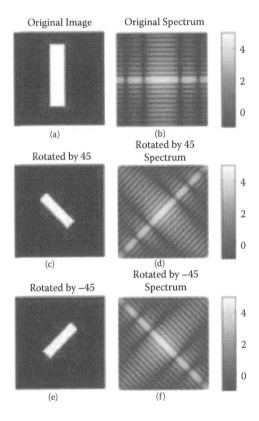

FIGURE 3.5

The rotation property of the Fourier spectrum. *Top row*: (a) Original image, (b) its Fourier spectrum. *Middle row*: (c) Rotated rectangle 45°, (d) corresponding spectrum. *Bottom row*: (e) Rotated rectangle by −45°, (f) corresponding spectrum. The third column indicates the intensity maps.

3.4.5 Distributive and Scaling

The Fourier transform (FT) and its inverse is distributive over the addition but not over the product:

$$\mathcal{F}\{f_1(x,y)+f_2(x,y)\} = \mathcal{F}\{f_1(x,y)\}+\mathcal{F}\{f_2(x,y)\}$$

$$\mathcal{F}\{f_1(x,y)\cdot f_2(x,y)\} \neq \mathcal{F}\{f_1(x,y)\}\cdot\mathcal{F}\{f_2(x,y)\}$$

For any a and b scalars, the scaling property of FT is as follows:

$$af(x,y) \leftrightarrow aF(u,v) \tag{3.21}$$

$$f(ax,by) \leftrightarrow \frac{1}{|ab|}\mathcal{F}\left(\frac{u}{a},\frac{v}{b}\right) \tag{3.22}$$

3.4.6 Average

The mean or average value of the 2D function is given as

$$\mu = \frac{1}{N^2}\sum_{x,y=0}^{N-1}\sum_{x,y=0}^{N-1} f(x,y)$$

$$\therefore F(0,0) = \frac{1}{N}\sum_{x,y=0}^{N-1}\sum_{x,y=0}^{N-1} f(x,y) \tag{3.23}$$

$$\frac{1}{N}F(0,0) = \mu$$

3.4.7 Convolution and Correlation

The convolution of the two functions $f(x,y)$ and $g(x,y)$ is denoted as $f(x,y)*g(x,y)$:

$$f(x,y)\cdot g(x,y) \leftrightarrow F(u,v)*G(u,v) \tag{3.24}$$

$$f(x,y)*g(x,y) \leftrightarrow F(u,v)\cdot G(u,v) \tag{3.25}$$

It can be seen that the convolution in time (frequency) domain leads to the product in the frequency (time) domain.

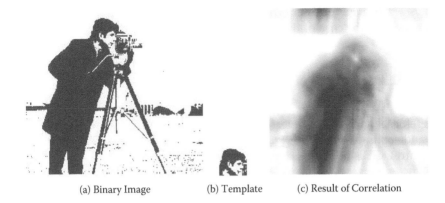

(a) Binary Image   (b) Template   (c) Result of Correlation

FIGURE 3.6
(a) Binary image, (b) template, (c) result of correlation.

3.4.8 Correlation

The correlation between the two functions $f(x,y)$ and $g(x,y)$ is denoted as $f(x,y) \circ g(x,y)$:

$$f(x,y) \circ g(x,y) \leftrightarrow F^*(u,v) \cdot G(u,v) \tag{3.26}$$

$$f^*(x,y) \cdot g(x,y) \leftrightarrow F(u,v) \circ G(u,v) \tag{3.27}$$

The example of correlation is shown in Figure 3.6. A binary image (Figure 3.6a) and a small template (Figure 3.6b) are correlated. The final correlation result is shown in Figure 3.6c. The highest correlation is seen as the brightest spot in the intensity domain. Using this property, the correlation can be used to identify the region of interest in the given image.

3.5 Two-Dimensional Discrete Cosine Transforms

The discrete cosine transform (DCT) expresses the given discrete signal as a linear combination of cosine functions at different frequencies with different amplitudes. DCT was introduced by Ahmed, Natarajan, and Rao in 1974. Consider the general structure for the transform as follows:

$$T(u,v) = \sum_{x,y=0}^{N-1} \sum_{x,y=0}^{N-1} f(x,y)\, g(x,y,u,v)$$

$$f(x,y) = \sum_{u,v,=0}^{N-1} \sum_{u,v=0}^{N-1} T(u,v) h(x,y,u,v)$$

where $g(:)$ and $h(:)$ represent the forward and the reverse kernel in the transformation. In the case of the DCT,

$$g(x,y,u,v) = \alpha(u)\alpha(v)\cos\left(\frac{(2x+1)u\pi}{2N}\right)\cos\left(\frac{(2y+1)v\pi}{2N}\right) = h(x,y,u,v)$$

Therefore, forward and reverse kernels are identical. They are also symmetric and separable. Then $\alpha(u)\alpha(v)$ defines the constant for DCT. Formally, the forward DCT is defined as

$$C(u,v) = \alpha(u)\alpha(v)\sum_{x,y=0}^{N-1}\sum_{x,y=0}^{N-1} f(x,y)\cos\left(\frac{(2x+1)u\pi}{2N}\right)$$

$$\times \cos\left(\frac{(2y+1)v\pi}{2N}\right); \quad 0 \leq u,v \leq N-1$$

(3.28)

Similarly, the inverse DCT is defined as

$$f(x,y) = \sum_{u,v,=0}^{N-1}\sum_{u,v=0}^{N-1} \alpha(u)\alpha(v)C(u,v)\cos\left(\frac{(2x+1)u\pi}{2N}\right)$$

$$\times \cos\left(\frac{(2y+1)v\pi}{2N}\right); \quad 0 \leq u,v \leq N-1$$

(3.29)

where

$$\alpha(u) = \frac{1}{\sqrt{N}}; \quad u = 0$$

and

$$\alpha(v) = \sqrt{\frac{2}{N}}; \quad u = 1,2,\ldots,N-1$$

The forward and the reverse kernels for $N = 2$ and $N = 4$ are shown in Figure 3.7a,b. The lighter-intensity values mean larger values of the kernel. Note that basis matrices show an increase in variation as one traverses from the top-left matrix ($u = 0$, $v = 0$) to the bottom-right matrix ($u = N - 1$, $v = N - 1$). It can be seen that the DCT basis is real and orthogonal, meaning $C = C^*$ and $C^{-1} = C^T$, where $*$ indicates the conjugate and T is the transpose operation, respectively.

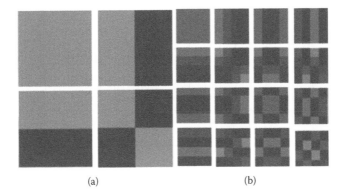

(a) (b)

FIGURE 3.7
(a) The DCT basis for N = 2, (b) the DCT basis for N = 4. The origin is considered to be the top-left corner.

The DCT is better at providing energy compaction. This means that a few DCT coefficients preserve a large proportion of the signal energy. Figure 3.8a shows a typical image, and its DCT spectrum is shown in Figure 3.8b. It can be observed that larger coefficients are present toward the top-left corner shown by the lighter intensity values. It can also be seen that most of the DCT coefficients have extremely small values indicated by the darker intensity values.

3.5.1 Comparison of Discrete Cosine Transforms and Discrete Fourier Transforms

The DCT is closely related to the DFT. In fact, the DCT can be computed using the fast Fourier transform (FFT). The relationship between them can be seen by considering the 1D sequence in Figure 3.9a. It is assumed that sequence is periodic with period N. The DFT considers that sequence outside this interval repeats as shown in Figure 3.9b. This implicit periodicity

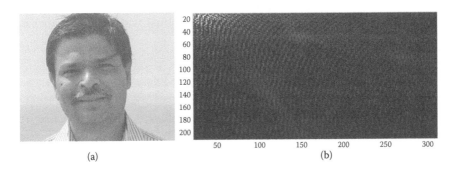

(a) (b)

FIGURE 3.8
(a) Typical image, (b) DCT spectrum.

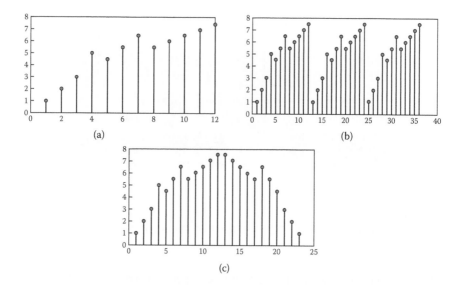

FIGURE 3.9
(a) One-dimensional sequence of length N (top row left), (b) Fourier transform view outside N (top row right), (c) DCT by mirroring one-dimensional sequence (bottom row).

assumption introduces sharp discontinuities at the starting and ending of the sequence which introduces high-frequency components. This effect is canceled out in the non-endpoint coefficients by the DFT, but when they are compressed, the Gibbs phenomenon causes blocking artifacts.

The DCT is obtained from the DFT by mirroring N point 1D sequence to get a $2N$ point sequence as shown in Figure 3.9c. Apply DFT to this $2N$ point sequence. DCT is then simply the first N points of the resulting DFT. As seen from Figure 3.9c, this mirroring does not introduce any discontinuity at the edges and thus the blocking of artifacts is avoided.

The DCT has numerous applications, but the most important one is in the domain of image compression. Specifically, the joint photographic expert group (JPEG) still image compression standard uses DCT as a transform. However, in this chapter we will explore a robust watermarking method based on DCT.

3.5.2 Application: DC–AC Coefficients Based on Multiple Watermarking Technique

Using DC coefficients for embedding a watermark is popular because the resulting method will be robust. This is due to the fact that (1) popular compression algorithms do not affect the DC coefficient and (2) DC coefficients have a very large amount of energy.

Traditionally, DC-coefficient-based methods are known to have a limited embedding capacity and poor perceptual quality for a watermarked image.

However, if the parameters are properly adjusted, these conflicting require-ments can be optimized. Most methods use a fixed step size during water-mark embedding. This results in analogous performance for all the images in spite of the fact that they are derived from different distributions. The method proposed in the next section uses a different step size for each image. The opti-mization between the robustness and embedding capacity can be achieved if the quantizer step size is adapted to image texture. Thus, one can increase robustness for fixed capacity by increasing step size.

The proposed method improves capacity by hiding in DC and AC coef-ficients simultaneously. Most prior works embedded 4096 bits in 512 × 512 images as against 8192 bits in the proposed work. This is done by embedding additional bits in the AC coefficients, which are strong enough to survive common image processing operations.

3.5.2.1 Deciding the Step Size

It is well known in the domain of data hiding that an image with higher tex-ture activity can hide a large number of bits without significant degradation in perceptual quality. It is difficult for the human visual system (HVS) to detect the changes in the region of high texture. Using this aspect, the first image shown in Figure 3.10a hides about 28,000 bits, and the image in Figure 3.10b hides about 7000 bits with peak signal-to-noise ratio (PSNR) greater than 50 dB.

For an image like that of a hurricane, one can trade robustness for capac-ity and perceptual degradation by increasing the step size. In quantization index modulation (QIM)-based techniques, increasing step size results in higher robustness at the cost of reduced perceptual quality. Reduced percep-tual quality can be compensated well if the image has high texture activi-ties. This notion is used in the current method to increase the step size if the image contains higher texture activity. The question to be addressed is how to detect the images with high texture activity quantitatively. Initially,

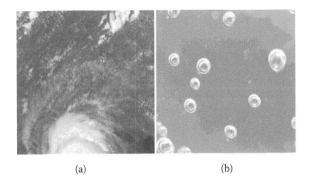

(a)   (b)

FIGURE 3.10
(a) Hurricane, (b) bubbles.

TABLE 3.1

Images and Their Entropy Values

Image	Entropy
Bubble	5.56
Building	6.44
Cameraman	7.04
Hurricane	7.26
Lena	7.32
Mandrill	7.29
Pirate	7.23

entropy was used as a measure for detecting an image with large texture changes. Table 3.1 indicates the entropy for some images.

It can be observed that the entropy can act as a measure for texture within an image and distinguish images like those of a hurricane and a bubble. Nevertheless, when the entropy does not have a large spread among values, it becomes difficult to segregate images, for example, images like those by Lena and Mandrill have very close entropy values but a large perceptual difference in textural activity. Thus, it becomes necessary to use other measures for deciding texture activity. Edge content within an image can be used to decide texture activity. High-pass filtering is performed on an image to extract edges. The resulting image is binary. The number of 1's in the binary image is counted to find a measure called the *edge count*. A higher value of edge count indicates large textural changes. Figure 3.11 shows high-pass filtered images of a bubble and a hurricane. Table 3.2 indicates the edge count for images.

It can be observed from Table 3.2 that the edge count covers a significantly large span of values. The high value of edge count indicates good texture activity, and vice versa. The proposed method uses three different values of step size based on the edge count. These values are shown in Table 3.3.

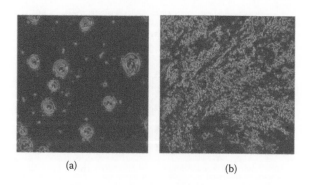

(a)   (b)

FIGURE 3.11
High-pass filtered images: (a) hurricane and (b) bubble.

TABLE 3.2

Edge Count for Images

Image	Edge Count
Bubble	5768
Building	14,081
Cameraman	14,930
Hurricane	24,384
Lena	13,865
Mandrill	31,374
Pirate	20,102

These values of step size are arrived at by the heuristic approach through experimentation for the given values of PSNR. These values of step size will be used in watermark embedding and extraction as described in the next section.

3.5.2.2 Watermark Embedding

- Divide the source image into nonoverlapping blocks of 8×8.
- Compute the forward DCT for each of these blocks.
- Scan the coefficients in a zigzag fashion.
- The DC coefficient and the first AC coefficient are used for watermark embedding in the block. These locations are shared between the encoder and decoder. The other coefficients in the block are left unaltered.
- With the 8192-random watermark bit (0 and 1) pattern with zero mean, the unit variance is generated.
- The watermark is embedded by rounding coefficients to even values if 0 is embedded and to odd values if 1 is embedded.
- The step size is obtained as a per edge count from Table 3.3, and is used for the QIM embedder.
- Apply the inverse DCT to obtain the watermarked image.

TABLE 3.3

Edge Counts and Step Sizes

Range of Edge Count	Step Size
<6000	10
6000 to 15,000	20
>15,000	30

3.5.2.3 Watermark Decoding

The decoder shares locations of potential coefficients in 8 × 8 blocks. The decoding algorithm is as follows:

- Divide the received watermarked image into nonoverlapping 8 × 8 blocks.
- Compute the forward DCT for each of these blocks.
- Scan the coefficients in a zigzag fashion.
- The DC coefficient and the first AC coefficient are used for watermark detection.
- These coefficients are rounded to the nearest integer level. If the integer is even, it is decoded as 0. If the integer is odd, then it is decoded as 1.
- The extracted bit pattern can be compared with the original to derive the method statistics.

3.5.2.4 Experimentation and Results

The algorithm is applied to 8-bit grayscale images with size 512 × 512. Figure 3.12 shows the test images that were chosen to represent different distributions.

Various test parameters at the encoder and the decoder are computed to evaluate the performance of the scheme. The parameters evaluated at the encoder are between the original and watermarked images. They are as follows:

1. Similarity factor (SF) or normalized correlation (NC)
2. Root mean square error (RMSE)

FIGURE 3.12
Test images.

TABLE 3.4

Encoder Parameters

NC	RMSE	DWR	PSNR
0.99	1.44	46.59	51.72

3. Document-to-watermark ratio (DWR) (in dB)
4. Peak signal-to-noise ratio (PSNR) (in dB)

For comparison with other methods, the parameters listed in Table 3.4 are derived for the Lena image with a step size of 20 and the number of bits embedded is 8192.

Various parameters evaluated at the decoder are as follows:

1. Watermark-to-noise ratio (WNR) (in dB)
2. Normalized correlation (NC)
3. Number of errors
4. Bit error rate (BER)
5. Signal-to-noise ratio (SNR) (in dB)
6. Root mean square error (RMSE)
7. Peak signal-to-noise ratio (PSNR) (in dB)

Parameters 1 to 4 are between a recovered and an original watermark. Parameters 5 to 7 are for a watermarked image that enters the communication channel, gets corrupted by noise within, and then appears out of it. Table 3.5 shows various decoder parameters for image Lena with 8192 bits recovered with various attacks.

TABLE 3.5

Decoder Parameters

WNR	NC	Bits Error	BER	SNR	RMSE	PSNR
JPEG Compression with a Quality Factor of 50						
58.3408	0.99	12	0.0015	31.76	3.03	44.31
Gaussian Noise N(0,0.0005)						
17.59	0.91	706	0.086	25.74	4.10	41.30
Salt and Pepper Noise (0.001)						
28.5093	0.97	237	0.028	46.07	1.48	51.46
Median Filter [3 × 3]						
23.1017	0.9504	407	0.0497	32.63	2.90	44.74

TABLE 3.6

Decoder Parameters for a Hurricane with a Step Size of 30

WNR	NC	Number of Bits in Error	BER	SNR	RMSE	PSNR
Inf	1	0	0	22.36	5.20	38.90

Table 3.6 shows various decoder parameters for the hurricane image with 8192 bits embedded and a JPEG compression with Q = 50 and step size of 30. This will increase the robustness with reduction in PSNR value as indicated in Table 3.6.

In Table 3.7, a comparison of the proposed work with other methods is shown in terms of PSNR and NC. A JPEG compression with Q = 50 and a step size of 20 is used as a benchmark for comparison among these methods.

In terms of other attacks, such as median filtering, with a size of 3×3, the comparison is illustrated in Table 3.8.

The effect of a JPEG compression on watermark recovery and other parameters is studied by subjecting the watermarked image to a JPEG compression with Q varying from 0 to 100. Figure 3.13a–d shows a variation of Q on parameters like WNR, NC, BER, and PSNR.

3.5.2.5 Conclusion and Discussion

This method demonstrates improved robustness with superior capacity than other existing works. The step size is chosen based on the textural content of an image. This will lead to increased robustness while compensating perceptual quality. However, the decision on step size value for a given image is taken heuristically. Future work can look into deriving this value in a more sophisticated mode.

The next section will look at 2D image transforms, which derive basis from the given data.

TABLE 3.7

Comparison among Different Methods with the Proposed DC-AC Coefficient Based Method

Method	Zeng and Qiu (2008)	Xiao and Wang (2008)	Proposed Method
PSNR	38.29	38	44.31
NC	0.95	0.94	0.99
Number of bits embedded	4096	4096	8192

TABLE 3.8

Comparison for Median Filtering 3×3

Method	Zeng and Qiu (2008)	Proposed Method
PSNR	38.6	44.74
NC	0.8913	0.95
Number of bits embedded	4096	8192

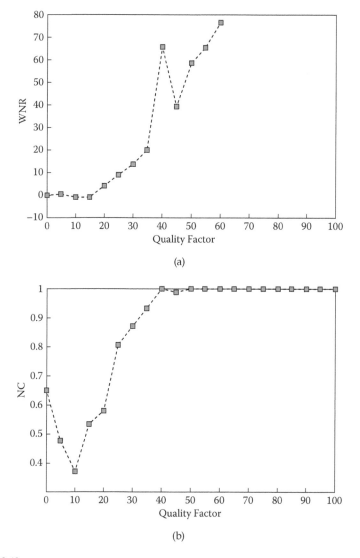

(a)

(b)

FIGURE 3.13
(a) Q versus WNR, (b) Q versus NC, (c) Q versus BER, and (d) Q versus PSNR. (*continued*)

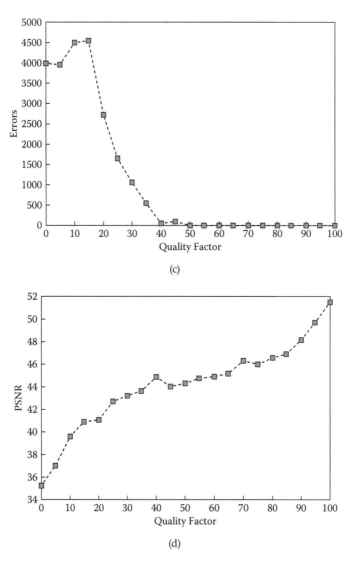

(c)

(d)

FIGURE 3.13 (*continued*)
(a) Q versus WNR, (b) Q versus NC, (c) Q versus BER, and (d) Q versus PSNR.

3.6 The Walsh-Hadamard Transform

A Walsh-Hadamard transform (WHT) is a nonsinusoidal transformation that decomposes the signal into a set of functions known as the *Walsh functions*. These functions are rectangular or square and orthogonal functions. It is a computationally simpler transform as it has no multipliers and has real values. Interestingly, Walsh functions take only two amplitude ±1. The 1D kernel for WHT is as follows:

$$g(x,u) = \frac{1}{N}(-1)^{\sum_i b_i(x)b_i(u)}, \quad i = 1,2,\ldots n-1$$

The forward WHT is then given as

$$\mathcal{H}(u) = \sum_{x=0}^{N-1} f(x)\frac{1}{N}(-1)^{\sum_i b_i(x)b_i(u)} \tag{3.30}$$

The inverse WHT is given as

$$f(x) = \sum_{u=0}^{N-1} \mathcal{H}(u)\frac{1}{N}(-1)^{\sum_i b_i(x)b_i(u)} \tag{3.31}$$

It should be noted that the forward and the reverse kernels for the WHT are identical. Like the case for 1D functions, the 2D forward and inverse kernels are given as follows:

$$g(x,y,u,v) = h(x,y,u,v) = \frac{1}{N}(-1)^{\sum_{i=0}^{n-1} b_i(x)p_i(u)+b_i(y)p_i(v)}$$

where $N = 2^n$. It should be noted that summation in the exponential term is done using modulo 2 arithmetic. Then $b_i(x)$ represents the *i*th bit from right to left in the binary representation of x. The p_i are computed using

$$p_0(u) = b_{n-1}(u)$$

$$p_1(u) = b_{n-1}(u) + b_{n-2}(u)$$

$$p_2(u) = b_{n-2}(u) + b_{n-3}(u)$$

$$\vdots$$

$$p_{n-1}(u) = b_1(u) + b_0(u)$$

where the sums are as per modulo 2 arithmetic. A typical Hadamard matrix for $N = 8$ is given as follows:

$$WH = \begin{bmatrix} 1 & 1 & 1 & 1 & 1 & 1 & 1 & 1 \\ 1 & -1 & 1 & -1 & 1 & -1 & 1 & -1 \\ 1 & 1 & -1 & -1 & 1 & 1 & -1 & -1 \\ 1 & -1 & -1 & 1 & 1 & -1 & -1 & 1 \\ 1 & 1 & 1 & 1 & -1 & -1 & -1 & -1 \\ 1 & -1 & 1 & -1 & -1 & 1 & -1 & 1 \\ 1 & 1 & -1 & -1 & -1 & -1 & 1 & 1 \\ 1 & -1 & -1 & 1 & -1 & 1 & 1 & -1 \end{bmatrix}$$

It is a symmetric matrix with columns containing the Walsh-Hadamard functions. Interestingly, there is no definite pattern of sign change along the column like we obtain in the cases of the DCT and DFT. The functions are not arranged in increasing order of sequences or number of zero crossings. Walsh functions along the columns in increasing order of sequences are as follows:

$$W = \begin{bmatrix} 1 & 1 & 1 & 1 & 1 & 1 & 1 & 1 \\ 1 & 1 & 1 & 1 & -1 & -1 & -1 & -1 \\ 1 & 1 & -1 & -1 & -1 & -1 & 1 & 1 \\ 1 & 1 & -1 & -1 & 1 & 1 & -1 & -1 \\ 1 & -1 & -1 & 1 & 1 & -1 & -1 & 1 \\ 1 & -1 & -1 & 1 & -1 & 1 & 1 & -1 \\ 1 & -1 & 1 & -1 & -1 & 1 & -1 & 1 \\ 1 & -1 & 1 & -1 & 1 & -1 & 1 & -1 \end{bmatrix}$$

This matrix reflects the change in sign along the columns. Frequency of sign change along the column increases as we move along the column.

WHT transform kernels are symmetric and separable. The Walsh-Hadamard basis function for $N = 4$ is shown in Figure 3.14.

The WH kernel consists of alternating ± 1. They are arranged in the checkerboard pattern. Each block has a 4×4 subblock. The white intensity level means +1 and black –1. For example, the top-left block has $u = v = 0$, and we get the values for $g(x, y, 0, 0); x, y = 0, 1, 2, 3$. It turns out that all values in this case are +1.

The WHT is simple to implement as the kernel takes the value of either +1 or –1. The WHT can be extensively used in filtering, speech processing, code division multiple access communication, medical signaling, nonlinear signal characterization and analysis, logical designing, and many more. However, one must note that it is a suboptimal transform from an energy

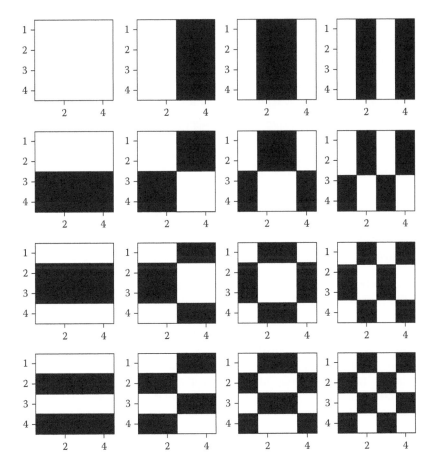

FIGURE 3.14
The Walsh basis for $N = 4$.

compaction perspective. So our quest for an optimal transform continues, which is explored further in the next section.

3.7 Optimal Transforms

A unitary transform has some desirable properties like the following:

1. *Energy compaction*—It can be shown that many unitary transforms pack a large proportion of total energy into a very few transform coefficients (refer to Figure 3.8b).

2. *De-correlation*—When it is applied to highly correlated input signals like images in a spatial domain, this results in uncorrelated output coefficients.

The covariance matrix can be used to measure the correlation among the different coefficients. The covariance matrix for transformation T is defined as

$$R_T \triangleq cov(T) = E\{(T - E\{T\})(T - E\{T\})^{*T}\} \tag{3.32}$$

After looking at some of the data-independent image transforms, we came across some interesting questions:

1. Is there an image transform with the best energy compaction?
2. Is such a transform capable of providing the highest degree of de-correlation?
3. Most importantly, can it be unitary?

The image transforms we have seen so far are data independent, meaning that the transform basis is fixed. They do not depend on the image signal being processed. For improving the image transform we need them to be *optimal* in the *statistical sense* so that it can be generalized for any type of signal. But how to define an optimality is also an interesting quest.

Assume that $x \in \mathbb{R}^N; E[x] = 0$. Then the covariance, in this case the autocorrelation matrix, is defined as $R_x = cov(x) = E[xx^H]$, where H indicates Hermitian. Then $R_x(i, j)$ will show the correlation between (x_i, x_j). And, R_x will be diag(:) matrix if all random variables of x are uncorrelated. Also R_x will be a symmetric matrix for real signals and Hermitian for complex signals (i.e., $R_x^T = R_x$, $R_x^H = R_x$). Moreover, R_x will be nonnegative definite meaning it will have real nonnegative eigenvalues.

Now apply a linear transformation $y = Ax$. The correlation matrix of y is

$$R_y = cov(y) = E[yy^H]$$

$$= E[Ax(Ax)^H] = E[Axx^H A^H]$$

$$= AE[xx^H]A^H = AR_x A^H \tag{3.33}$$

The transform matrix A has the following structure:

$$A = \begin{bmatrix} a_0 \\ a_1 \\ a_2 \\ \vdots \\ a_{N-1} \end{bmatrix}$$

and it has a property that $a_i a_i^* = 1$ and $a_i a_j^* = 0; i \neq j$.

This will allow $\|y\|^2 = E[y^H y] = \|x\|^2$, where $\|\cdot\|$ is a norm:

$$\|y\|^2 = \sum_{i=0}^{N-1} y_i^2$$

Now the problem for optimal transform can be formulated as follows:

$$max.E[y^H y]$$

$$\text{s.t. } y_i = a_i x, a_i a_i^* = 1, \quad a_i a_j^* = 0; \quad i \neq j$$

Without going into the details of the proof for the above maximization problem, it can be shown that the energy for y can be maximized if a_i^* are the eigenvectors of R_x; $R_x a_i^* = \lambda_i a_i^*$ (i.e., transformation matrix A is formed using the eigenvectors of the covariance matrix).

3.8 The Karhunen-Loeve Transform: Two-Dimensional Image Transform with a Data-Dependent Basis

The Karhunen-Loeve transform (KLT) is also an example of the unitary transform with basis vectors in A that are orthonormal eigenvectors of R_x. The x is a vectorized representation of the image. The forward KLT is given as

$$y = Ax \tag{3.34}$$

$$\text{with} \quad A = \mathbb{R}^{N \times N}, \ A = \left[a_0^T, a_1^T, a_2^T \dots, a_{N-1}^T \right],$$
$$\text{where} \quad R_x a_i = \lambda_i a_i; i = 0, 1, \dots N-1$$

It can be seen that kernel A in the KLT is derived from the data, and it operates on the statistical characteristics of the vector representation of the image. The inverse KLT is simply given as

$$x = A^{-1} y = A^H y \tag{3.35}$$

3.8.1 Properties of the Karhunen-Loeve Transform

3.8.1.1 De-Correlation

As per Equation (3.33):

$$R_y = \begin{bmatrix} \lambda_0 & \cdots & 0 & 0 \\ \vdots & \lambda_1 & \vdots & 0 \\ 0 & \vdots & \ddots & 0 \\ 0 & 0 & 0 & \lambda_{N-1} \end{bmatrix} = diag(\lambda_k); k = 0, \dots, N-1$$

Moreover, λ's for $R_y = R_x$. It shows R_y is a diagonal matrix when all its off-diagonal elements are zero. This means that elements of y are uncorrelated with each other. KLT establishes a new coordinate system with the origin at the centroid of the signal. The axis of the new coordinate system will be parallel to the eigenvectors of R_x. Thus, KLT is typically a rotational transform that aligns the data with the eigenvectors.

3.8.1.2 Minimizing the Mean Square Error with a Limited Basis

Signal x is transformed using A, then the number of coefficients in y is restricted to M, where $1 \le M \le N$ (Figure 3.15). This is done by using

$$
I_M = \begin{bmatrix} 1 & 0 & \cdots & 0 \\ \vdots & \ddots & \ddots & \vdots \\ 0 & 0 & 1 & 0 \\ 0 & 0 & 0 & 0 \end{bmatrix}
$$

Coefficients related to the eigenvectors corresponding to the first M highest eigenvalues are preserved. Due to use of the limited basis, the reconstructed signal will have an error. The reconstruction error is quantified as follows:

$$
\mathbb{E} \triangleq \frac{1}{N} E \left[\sum_{i=0}^{N-1} (x_i - \hat{x}_i)^2 \right]
$$

$$
\frac{1}{N} Tr \left\{ E \left[(x - \hat{x})(x - \hat{x})^{*H} \right] \right\}
$$

where $Tr\{\cdot\}$ is the trace of a matrix:

$$
\mathbb{E} = \sum_{i=0}^{N-1} \lambda_i - \sum_{j=0}^{M-1} \lambda_j = \sum_{l=M+1}^{N} \lambda_l
$$

So the reconstruction error is equal to the sum of the smallest eigenvalues, and therefore, the mean square error (MSE) is minimized. The KL transform is an optimal transform as it minimized the MSE.

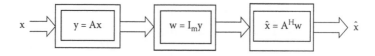

FIGURE 3.15
The basis restraint in the KL transform.

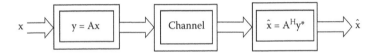

x ⟹ | y = Ax | ⟹ | Channel | ⟹ | $\hat{x} = A^H y^*$ | ⟹ $\hat{x}$

FIGURE 3.16
The KL transformed signal traveling through a noisy channel.

3.8.1.3 *Minimizing the Transmission Rate at a Given Noise*

The signal x is traveling through a noisy channel and transformation $y \rightarrow y^*$ (Figure 3.16). The noise specification is known or can be estimated as

$$n = \frac{1}{N} E[(x - \hat{x})^{*H}(x - \hat{x})]$$

The KLT among all the unitary transforms achieves the minimum transmission rate $R(A) \leq R(\varphi)$.

Thus, it can be summarized that KLT achieves minimum MSE and decorrelates the samples optimally. However, it is not commonly used in image compression due to its computational requirements. It needs a very good estimator for second-order statistics. Assuming $x \in \mathbb{R}^N$, the computation of $R_x \sim \mathcal{O}(N^2)$, the eigenvalue of R_x takes $\mathcal{O}(N^3)$ and linear transformation $\mathcal{O}(N^2)$. Even with a faster transform the computation linear operation will be $\mathcal{O}(N \log N)$. Last, we evaluate the performance of the KLT. Figure 3.17a–f

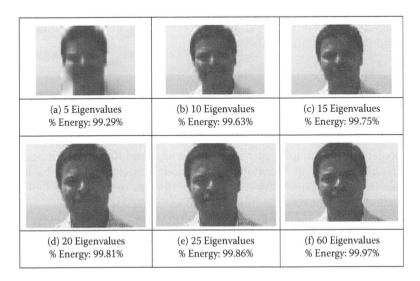

| (a) 5 Eigenvalues % Energy: 99.29% | (b) 10 Eigenvalues % Energy: 99.63% | (c) 15 Eigenvalues % Energy: 99.75% |
| (d) 20 Eigenvalues % Energy: 99.81% | (e) 25 Eigenvalues % Energy: 99.86% | (f) 60 Eigenvalues % Energy: 99.97% |

FIGURE 3.17
An image reconstruction using different eigenvalues and the percentage of total energy contained in them.

shows progressive image reconstruction using 5 to 60 of the highest eigen-values. The image perceptual fidelity improves as more eigenvalues are retained. However, it is very interesting to note that the five largest eigenvalues preserve 99.29% (Figure 3.17a) of the total energy.

Qualitatively, the comparison of three transforms DCT, DFT, and KLT on various parameters is as shown in the following table.

Parameter	DFT	DCT	KLT
Lossless inverse transform	Yes	Yes	Yes
Energy preservation in transform domain as per Parseval's theorem	Yes	Yes	Yes
Energy compaction	No	Close to KLT for first-order stationary Markov source	Optimal
Orthonormal basis complete over a set	Yes	Yes	Yes
Signal-dependent basis function	No	No	Yes
Real valued coefficients	No	Yes	Yes
Separability	Yes	Yes	No
Similar forward and reverse kernels	Yes with change in phase value	Yes	No
Faster implementation using structure like butterfly	Yes	Yes	No

We can summarize that the DCT is a close approximation to the KLT, at least for the first-order Markov process. The DCT is optimal for a highly correlated signal; it does not depend on the specific data, and the fast algorithm for computing transform is also available.

3.9 Summary

The content of this chapter provides the background for many applications. The treatment of image transforms even though brief provides a basic idea of their individual capabilities. The discussion on data-independent basis and data-dependent basis is fundamental in understanding many applications discussed in this book, specifically transform coding used in image compression and the rise of compressive sensing theory based on sparsification done by these transforms. While looking at the performance of various image transforms, it can be seen from Figure 3.18 that DCT is similar to the KL transform in providing the energy compaction. However, the limitation

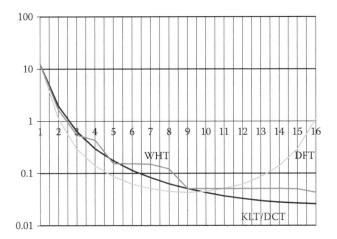

FIGURE 3.18
The distribution of variance σ^2 of transform coefficients for a typical first-order stationary Markov process.

of these transforms is that they cannot achieve simultaneous time and frequency localization. If the signals are stationary, they provide complete information about the spectral content, but for nonstationary signals they fail to provide information about the time at which a particular frequency component occurred in the signal. This time frequency localization is achieved by wavelet transform, which is discussed in this book separately.

Bibliography

N. Ahmed, T. Natarajan, and K. R. Rao, Discrete cosine transform, *IEEE Transactions on Computers*, C-23: 90–93, January 1974.

N. Ahmed and K. R. Rao, *Orthogonal Transform for Digital Signal Processing*, Springer Verlag, New York, 1975.

H. C. Andrews and K. Casparie, *Orthogonal Transformations, Computer Techniques in Image Processing*, Academic Press, New York, 1970.

K. G. Beauchamp, *Applications of Walsh and Related Functions—With an Introduction to Sequency Theory*, Academic Press, New York, 1984.

T. Beer, Walsh transforms, *American Journal of Phyics*, 49(5): 466, May 1981.

V. Bhaskaran and K. Konstantinos, *Image and Video Compression Standards: Algorithm and Architecture*, Kluwer, Boston, MA, 1997.

R. M. Bracewell, *The Fourier Transform and Its Application*, McGraw-Hill, New York, 1968.

K. R. Castleman, *Digital Image Processing*, 2/e, Prentice Hall, Upper Saddle River, NJ, 1996.

D. C. Champeney, *A Handbook of Fourier Theorems*, Cambridge University Press, New York, 1987.

Chanda and D. Dutta Majumder, *Digital Image Processing and Analysis*, EEE, Prentice Hall, New Delhi, India, 2009.

J. W. Cooley and J. W. Tukey, An algorithm for the machine computation of complex Fourier series, *Mathematics of Computation*, 19: 297–301, 1965.

H. Enomoto and K. Shibata, Orthogonal transform coding system for television signals, *IEEE Trans. on Electromagnetic Compatibility*, 13: 11–17, 1971.

W. M. Gentleman, Matrix multiplication and fast Fourier transformations, *Bell System Technical Journal*, 47: 1099–1103, 1968.

R. C. Gonzalez and R. E. Woods, *Digital Image Processing*, 3e, Pearson Education, New York, 2009.

H. Hotelling, Analysis of complex statistical variable into principal components, *Journal of Educational Psychology*, 24: 417–441, 498–520, 1933.

A. K. Jain, A fast Karhunen-Loeve transform for finite discrete images, in *Proc. National Electronics Conference*, pp. 323–328, Chicago, IL, October 1974.

A. K. Jain, A sinusoidal family of unitary transforms, *IEEE Trans. Pattern Analysis and Machine Intelligence*, 1: 356–365, 1979.

A. K. Jain, *Fundamentals of Digital Image Processing*, Prentice Hall, Englewood Cliffs, NJ, 1989.

N. S. Jayant and P. Noll, *Digital Coding of Waveforms*, Prentice Hall, Englewood Cliffs, New Jersey, 1984.

M. A. Joshi, *Digital Image Processing: An Algorithmic Approach*, EEE, Prentice Hall, New Delhi, India, 2009.

M. V. Joshi, V. Joshi, and M. S. Raval, Multilevel semi-fragile watermarking technique for improving biometric fingerprint system security, in *Intelligent Interactive Technologies and Multimedia*, A. Agrawal, R. C. Tripathi, E. Yi-Luen Do, and M. D. Tiwari, Eds., Springer, Heidelberg, 2013, pp. 272–283.

H. Karhunen, On linear methods in probability theory, *Tech. Report Doc, T-131*, Rand Corporation, 1947.

M. J. Lighthill, *Introduction to Fourier Analysis and Generalized Function*, Cambridge University Press, New York, 1960.

M. Loeve, *Fonctions Aleatoires de Seconde Ordre*, Hermann, Paris, 1948.

W. B. Pennebaker and J. L. Mitchell, *JPEG Still Image Data Compression Standard*, Van Nostrand Reinhold, New York, 1993.

A. Popoulis, *The Fourier Integral and Its Applications*, McGraw-Hill, New York, 1962.

W. K. Pratt, *Digital Image Processing*, 2/e, Wiley Interscience, New York, 1991.

W. K. Pratt, H. C. Andrews, and J. Kane, Hadamard transform image coding, *Proc. IEEE*, 57: 58–68, 1969.

C. R. Rao and S. K. Mitra, *Generalised Inverse of Matrices and Its Applications*, Wiley, New York, 1971.

M. S. Raval, M. V. Joshi, and S. Kher, Fuzzy neural based copyright protection scheme for superresolution, in *IEEE Int. Conf. on Sys., Man and Cyber. (SMC 2013)*, pp. 328–332, Manchester, UK, October 13–16, 2013.

J. O. Smith III, *Mathematics of Discrete Fourier Transform*, W3K, CCRMA, Stanford, CA, 2003.

J. L. Walsh, A closed set of orthogonal functions, *American Journal of Mathematics*, 55: 5–24, 1923.

J. Xiao and Y. Wang, Toward a better understanding of DCT coefficients in watermarking, *Proc. IEEE Pacific-Asia Workshop on Computational Intelligence and Industrial Application*, pp. 206–209, 2008.

G. Zeng and Z. Qiu, Image watermarking based on DC component in DCT, *Proc. International Symposium on Intelligent Information Technology Application Workshops*, pp. 573–576, 2008.

4

Wavelet-Based Image Compression

4.1 Introduction

In Chapter 3, we studied how the discrete cosine transform (DCT) is used for the compression of images. The basic properties of getting a higher compression ratio include energy compaction and the de-correlation property of the DCT. The JPEG lossy image compression standard that uses DCT suffers through artifacts like blockiness and ringing, which degrade the quality of the decompressed image.

In this chapter, we are going to study another transform called the *wavelet transform*, which possesses properties like energy compaction and de-correlation and is used in image compression, reducing an effect of blockiness artifact. The wavelet is the most popular transform in the field of signal processing and is used in a number of applications. The concept of wavelets, continuous and discrete wavelets and their properties, and wavelet-based multiresolution analysis (MRA), which is used for image compression, are described in this chapter. The wavelet-based compression is used in JPEG 2000 or the J2K image compression standard.

4.2 The Short-Time Fourier Transform

The Fourier transform has long been used for signal analysis. The Fourier transform provides information about the frequency spectrum of the signal. It presents the frequencies and their amplitudes present in the signal. The Fourier transform only presents the signal frequencies and not the time instance at which a particular frequency occurs. Another drawback of the Fourier transform is that it works better with stationary signals.

This frequency localization problem is overcome by the short-time Fourier transform (STFT) in which the signal is analyzed in a particular time interval, taking the Fourier transform in that interval. For analysis of the low-frequency signal, the time interval should be large, and for a high-frequency

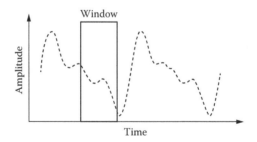

FIGURE 4.1
Short-time Fourier transform.

signal it should be small. Thus, for decomposing, the time interval (scale) must be varied. The process is as shown in Figure 4.1.

The Fourier transform of the windowed signal is given by

$$F(\omega, \tau) = \int_{-\infty}^{+\infty} f(t) w(t - \tau) e^{-j\omega t} \, dt \tag{4.1}$$

where $w(t)$ is a time-domain windowing function.

The value τ indicates the starting position of the window. A function $f(t)$ is mapped into a two-dimensional plane (ω, τ) by the STFT. If the window function is Gaussian, then the STFT is often called a *Gabor transform*. From the STFT, we can only know which frequency bands exist at what time intervals and cannot know exactly what frequency exists at what time instance. This problem is known as the *resolution problem*. The STFT is related to the width of the window function used. The narrower the window width is, the poorer the frequency resolution is, and vice versa. This is possible if the window with different widths is translated over the signal and frequencies are determined. The solution is wavelets, the wavelet transform.

4.3 Wavelets

Wavelet means "small wave." It overcomes the resolution problem and can be used as an alternative approach to the STFT. It is similar to STFT in which the signal is multiplied with a wavelet function. The Fourier transform of the signal under the window is not taken. As the transform is computed for each spectral component, the width of the window gets changed. This is nothing but multiresolution analysis (MRA), with which the signal is decomposed into several subbands of high and low frequencies at different scales of wavelets. The transformation used for the analog signal is the continuous wavelet transform (CWT), and for discrete signals, the discrete wavelet transform (DWT). Discrete transformation is widely used and easily implemented using computers.

4.4 The Continuous Wavelet Transform

Let $f(t)$ be any square integrable function. The CWT of $f(t)$ with respect to a wavelet function $\psi(t)$ is defined as

$$W_{(a,b)} = \int_{-\infty}^{+\infty} f(t)\frac{1}{\sqrt{a}}\psi\left(t - \frac{b}{a}\right)dt \tag{4.2}$$

where a and b are real, a is scale, b is translation (time), and $W_{(a,b)}$ are the wavelet coefficients generated at every scale and different translation. Then, $\psi(t)$ is the mother wavelet function. If the wavelet function matches with the signal, then the magnitude of the coefficients will be higher.

The mother wavelet function is translated over the signal with different scales. For $a > 1$, the function is dilated, and for $a < 1$ it is compressed. The dilated function will provide the low-frequency content of the signal, and the compressed function will provide the high-frequency content. Low and high frequencies are often called *approximations* and *details*, respectively. The CWT can be considered as the inner product of the signal with a basis (wavelet) function.

The mother wavelet function should satisfy the following properties:

1. $\psi(t)$ is the square integrable function:

$$\int_{-\infty}^{\infty} |\psi(t)|^2 dt < \infty \tag{4.3}$$

2. Energy of the signal $\psi\,a,b(t)$ is the same for all a, b:

$$\int_{-\infty}^{\infty} |\psi_{a,b}(t)|^2 dt = \int_{-\infty}^{\infty} |\psi(t)|^2 dt \tag{4.4}$$

3. Wavelets satisfy admissibility conditions:

$$\int_{-\infty}^{\infty} |\psi(\omega)|^2 /|\omega|\, d\omega < +\infty \tag{4.5}$$

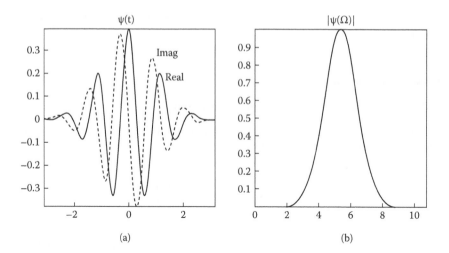

FIGURE 4.2
(a) Oscillatory wavelet function $\psi(t)$; (b) $\psi(\omega)$ bandpass spectrum of wavelet.

4. $\psi(t)$ is oscillatory:

$$\int\limits_{-\infty}^{\infty} |\psi(t)|\, dt = 0 \tag{4.6}$$

5. $\|\psi(\omega)|^2|_{\omega=0} = 0$ (i.e., the wavelet shows a bandpass-like spectrum)

The oscillatory wavelet function is as shown in Figure 4.2. Equation (4.2) can also be written as

$$w_{a,b} = \int\limits_{-\infty}^{\infty} f(t)\frac{1}{\sqrt{a}}\psi(a,b) \tag{4.7}$$

where

$$\psi(a,b) = \frac{1}{\sqrt{a}}\psi\left(t - \frac{b}{a}\right) \tag{4.8}$$

Following are a few more details about wavelets.

The wavelet transform $w(a,b)$ can be considered as the output of the filter with impulse response $\psi_a(-b)$ and the input $f(t)$ for given a. The filter defined by wavelet function is a bandpass filter. Scaling factor a shifts the center frequency of the filter $1/a$ times. Bandwidth also changes $1/a$ times. Thus, the bandpass filter defined is a constant Q filter. To localize the event in the frequency domain, the bandwidth of the filter should be small, but this increases uncertainty in the time window, while a high-bandwidth filter is more localized in the time domain. By using different values of a and b, an event can be localized in the time domain and the frequency domain.

Smaller values of a shift the bandpass filter center frequency to a higher side, and this provides details or small-scale information. While large values shift the center frequency to lower values, this gives low-frequency contents or coarse information.

4.5 Inverse Continuous Wavelet Transform

The original signal can be recovered as

$$f(t) = \frac{1}{c} \int\limits_{-\infty}^{\infty} \int\limits_{-\infty}^{\infty} \frac{1}{|a|^2} \, w(a,b) \left| \psi_{a,b}(t) \right| dadb \tag{4.9}$$

where

$$c = \int\limits_{-\infty}^{\infty} \frac{|\psi(\omega)|^2}{\omega} d\omega \tag{4.10}$$

For the existence of the integral, c needs to be finite. And for finite c (constant), $\psi(0) = 0$ is a condition that should be met. The value of $\psi(0)$ is the average value of $\psi(t)$; hence, the mother wavelet requires zero mean. The state with finite c is known as the admissibility condition.

For the wavelets to have finite energy, $\psi(t)$ has to decay as ω tends to infinity. The necessity of these requirements indicates that there is a concentration of energy in a narrow frequency band in ψ, and this gives the wavelet its frequency localization capability.

4.6 The Discrete Wavelet Transform

The CWT maps the one-dimensional signal $f(t)$ to a function $w(a,b)$ of two continuous variables a and b. Variable a is the dilation, and b is the translational parameter. The region of support is the region for which $w(a,b) \neq 0$. A redundant representation of the signal is indicated by the region of support of CWT. The entire support of $w(a,b)$ is not required to recover $f(t)$. A nonredundant wavelet representation is as follows:

$$f(t) = \sum_{k=-\infty}^{k=\infty} d(k,l) 2^{-k/2} \psi(2^{-k} t - l) \tag{4.11}$$

Here, $a = 2^k$ dilation takes place for the discrete values; and $d(k,l)$ is the wavelet transform of $f(t)$ at $a = 2^k, b = 2^k l$. This is equivalent to sampling the coordinates (a,b) at the intervals differing by a factor of two. Such sampling is known as dyadic sampling, and $d(k,l)$, a two-dimensional sequence, is referred to as the discrete wavelet transform (DWT) of $f(t)$.

4.6.1 Wavelet Transform—Multiresolution Analysis

An N-dimensional data sequence X_n can be considered as an N-dimensional vector in N-dimensional space V_n. If we represent the same vector using the $N - 1$ dimensional subspace we get an approximation of the signal X_{n-1} (i.e., we are representing the signal with lower resolution). The error between the actual signal and approximation is

$$e_n = X_n - X_{n-1} \tag{4.12}$$

This logic can be extended so that

$$X_n = X_0 + e_1 + e_2 + \cdots + e_{n-1} \tag{4.13}$$

This becomes the multiresolution decomposition of the signal. Space V_n can be viewed as the union of subspaces $V_1, V_2, V_3, \ldots V_{n-1}$. Projection of vector X_n in subspace X_k is the representation of X_n at a lower scale or lower precision.

Thus, X_0 is the coarsest representation, and $e_1, e_2, \ldots e_n$ are the details. In subspace $V_{j+1} = V_j \oplus W_j$, where W_j is called the detail space at resolution level j and is orthogonal to V_j. This means that the inner product between any element in W_j and any element in V_j vanishes. We can continue the decomposition of the V space and obtain

$$V_{j+1} = W_j \oplus V_j = W_j \oplus W_{j-1} \oplus V_{j-1} \tag{4.14}$$

4.6.1.1 Scaling Function

The function ϕ, which generates the basis functions for all the spaces $\{V_j\}$, is called the *scaling function* of the MRA:

$$\phi_{j,k}(t) = 2^{j/2}\phi(2^j t - k) \tag{4.15}$$

The scaling function should satisfy the following properties:

1. The scaling function is orthogonal to its integer translation.
2. The subspaces spanned by the scaling function at low scales are nested within higher scales.

3. The only function that is common to all subspaces is $f(x) = 0$.

4. Any function can be represented within arbitrary precision.

The expansion function of subspace V_j can be represented as the weighted sum of the scaling function of subspace V_{j+1}:

$$\phi_{j,k} = \sum \phi_{j+1}(x) \tag{4.16}$$

4.6.1.2 Wavelet Function

Let wavelet function $\psi(x)$ be such that along with its integer translates and binary scaling it spans the difference between any two adjacent scaling subspaces V_j and V_{j+1}:

$$W1 = V_2 - V_1 \tag{4.17}$$

$$W_j = V_{j+1} - V_j = 2^{j/2}\psi(2^j x - k) \tag{4.18}$$

The space of all measurable integrable function is as follows:

$$L^2(R) = V_0 \oplus W_0 \oplus W_1 \oplus .... \tag{4.19}$$

Since the wavelet space resides within the spaces spanned by the next higher resolution scaling function, it can be expressed as the weighted sum of the shifted double-resolution scaling function:

$$f(x) = \sum_{j=0}^{\infty} a_{j0,k}\phi_{j0,k}f(x) + \sum_{j=0}^{\infty} b_{j0,k}\psi_{j,k}f(x) \tag{4.20}$$

where $a_{j0}k$ is the approximation coefficient, and $b_{j0}k$ is the detail coefficient.

$$a_{j0,k} = \sum_{m=-\infty}^{m=\infty} a_{j-1,k}\frac{c(m-2k)}{2} \tag{4.21}$$

$$b_{j0,k} = \sum_{m=-\infty}^{m=\infty} a_{j-1,k}\frac{d(m-2k)}{2} \tag{4.22}$$

where

$$\frac{c(m)}{2} = < \phi(t), \phi(2t - m) >$$ (4.23)

$$\frac{d(m)}{2} = < \psi(t), \phi(2t - m) >$$ (4.24)

Let the filter

$$h(k) = \frac{c(k)}{2}$$ (4.25)

$$g(k) = \frac{d(k)}{2}$$ (4.26)

$$a_{j,k} = \sum_{-\infty}^{\infty} a_{j-1,k} \frac{h(2k - m)}{2}$$ (4.27)

$$b_{j,k} = \sum_{m=-\infty}^{\infty} a_{j-1,k} \frac{g(2k - m)}{2}$$ (4.28)

where $a(j,k)$ is obtained from $a(j - 1,k)$ by convolving later with the sequence $h(k)$ and retaining the even index samples. Even index samples are obtained by decimating the output of filter $h(k)$ $b(j,k)$, which is obtained from $a(j - 1,k)$ by convolving later with the sequence $g(k)$ and retaining the even index samples. Even index samples are obtained by decimating the output of filter $h(k)$.

Thus, wavelet coefficients can be obtained by filtering.

4.6.2 Properties of the Digital Filter

$$\sum_{n} h(n) = 1$$ (4.29)

$$H(\omega) = \sum_{n} h(n) e^{-j\omega n}$$ (4.30)

$$H(0) = 1$$ (4.31)

that is, a low-pass filter (LPF).

$$|H(\omega)|^2 + |H(\omega + \Pi)|^2 = 1$$ (4.32)

It is a high-pass filter (HPF) and shown in Figure 4.3.

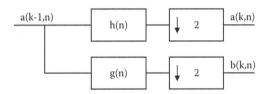

FIGURE 4.3
Analysis filter bank.

Wavelet transform of the discrete time signal is called DTWT. It can be found by using a filter bank similar to the continuous time signal.

4.6.3 Two-Dimensional Wavelet

To use wavelets for image compression, the rows and columns of the image must be scanned, and the wavelet coefficients need to be generated. The two-dimensional wavelet is presented in Figure 4.5 with the coefficients generated after applying to the image:

$$f(x,y) = \sum_{n=-\infty}^{\infty} \sum_{p=-\infty}^{\infty} a_0(n,p)\phi(x-n)\phi(y-p) + \sum_{n=-\infty}^{\infty} \sum_{p=-\infty}^{\infty} b_0(n,p)\phi(x-n)\psi(y-p)$$

$$+ \sum_{n=-\infty}^{\infty} \sum_{p=-\infty}^{\infty} c_0(n,p)\psi(x-n)\phi(y-p) + \sum_{n=-\infty}^{\infty} \sum_{p=-\infty}^{\infty} d_0(n,p)\psi(x-n)\psi(y-p)$$

$$(4.33)$$

The following are a few basic points:

- We can change the resolution of the signal, that is, the amount of coarse and detailed information within the signal, by the filtering operations and the scale by upsampling and downsampling operations.

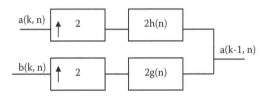

FIGURE 4.4
Reconstruction (synthesis) filter bank.

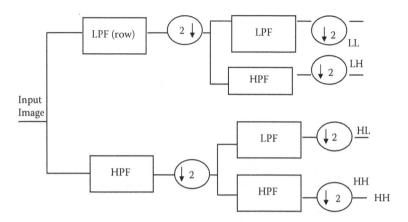

FIGURE 4.5
Two-dimensional wavelet decomposition.

- Multiresolution analysis (MRA) results from the implementation of the filter bank of the wavelet transform.
- Analysis of the signal at different frequency bands with different resolutions is performed by the DWT by decomposing the signal into low-frequency approximation and high-frequency detailed information.
- DWT operates with two functions, called scaling and wavelet functions, which are used in low- and high-pass filtering, respectively.
- If the detail coefficients are thresholded or made to be zero and inverse DWT is taken, the reconstructed signal contains fewer high-frequency components. This approach is used for removing noise, which happens to be the high frequency. This process is called *denoising*.

The multilevel decomposition of the signal is shown in Figure 4.6.

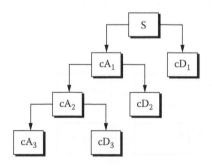

FIGURE 4.6
Multilevel signal decomposition using filters. (Image courtesy of MATLAB Toolbox.)

4.7 Wavelet Families

In a practical way, wavelet decomposition is done using wavelet filters. Varieties of wavelet filters are available, which are grouped under different families (e.g., Haar, Daubechies, Coeiflet, Morlet, Mayer, etc.). These wavelet filters are based on the wavelet function, which satisfies all criteria.

The type of filter, its length, and the number of decomposition stages should be considered when the wavelet transform is used for decomposition.

All the wavelets are theoretically reversible, but due to floating point arithmetic used in any processor and precision limitation, wavelet transform and its inverse provide very slightly different values of inversed samples.

The Haar, which is the basic wavelet, and its scaling and wavelet functions are shown in Figure 4.7.

Daubechies has a complete family of wavelets starting from 1, 2, ... 45 and is presented as db1,db2,db3,db4, ... and so on. The scaling and wavelet function of db4 and db8 are shown in Figure 4.8.

4.8 Choice of Wavelet Function

The choice of wavelet filter is an important parameter for any compression scheme.

Important properties of wavelet functions in image compression applications are as follows:

1. Compact support (leads to efficient implementation)
2. Symmetry (useful in avoiding dephasing in image processing)
3. Orthogonality (allows fast algorithm)
4. Regularity and degree of smoothness (related to filter order or filter length)

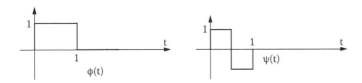

FIGURE 4.7
Haar scaling and wavelet function.

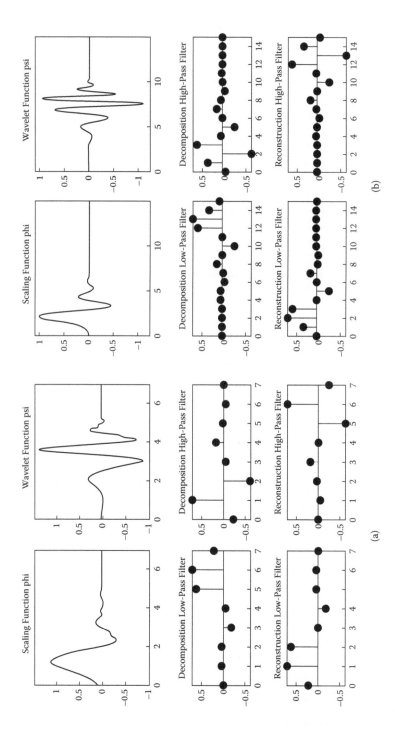

FIGURE 4.8
(a) db4; (b) db8 wavelet scaling and wavelet functions. (Image courtesy of MATLAB Toolbox.)

4.9 Discrete Wavelet Transform–Based Image Compression

Wavelets are widely used in compressing images. Wavelets exhibit a high energy compaction property and de-correlation like DCT, due to which quantization of the wavelet coefficients is possible. Wavelet-based compression is used in the JPEG 2000 (J2K) image compression standard.

In the DCT, one complicated dilemma is choosing the appropriate width of the block in the window in order to obtain the best models for the behavior of the image signals. Wavelet transform provides an elegant framework in which statistical behaviors (smooth and high frequency) of image data can be analyzed with equal importance. This facilitates a good trade-off between frequency and spatial domain with the wavelet representation of the image data. The approach for compressing images using wavelets is shown by the block diagram in Figure 4.9.

As discussed in the previous section, an image is filtered (decomposed) into approximation and detail. In Figure 4.9, with two dimensions, scanned row and column-wise, we get one approximation (LL1) and three details: horizontal (HL1), vertical (LH1), and diagonal (HH1). For next-level decomposition, an approximation is again filtered to divide into four more, thus getting to the seventh level. The approach for decomposition is shown in Figure 4.10.

For compression, wavelet coefficients are quantized by the thresholding of the wavelet coefficients. Thresholds can be different for different subbands depending on their significance, which mostly depends on the variance or energy. A large size for the steps of quantization can be used to quantize the low-energy subbands.

The following MATLAB® program shows how a very basic-level compression can be done using wavelets. MATLAB has a very good collection of

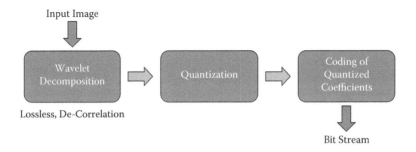

FIGURE 4.9
Block diagram of wavelet-based image compression.

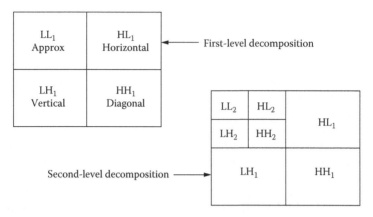

FIGURE 4.10
Approach for decomposition of image into subbands.

wavelet functions in its WAVELET Toolbox. Also the decomposition can be observed using a function called WAVEMENU:

```
%% Program for Compression of Image using Wavelets.
clear all;
clc;
close all;
I = imread('peppers.tif');
I = rgb2gray(I);
figure;
imshow(I);
I = double(I);
[cA1,cH1,cV1,cD1] = dwt2(I,'db1');
WI = [cA1 cH1 ;cV1 cD1];
figure;
imshow(uint8(WI));
%cA1t = wthresh(cA1,'h',400);
cV1t = wthresh(cV1,'h',20);
cH1t = wthresh(cH1,'h',20);
cD1t = wthresh(cD1,'h',20);
B = idwt2(cA1,cH1t,cV1t,cD1t,'db1');
figure;
imshow(uint8(B));
%%%%%%%%%
x = double(I);
y = double(B);
diff = x-y;            % difference of two images
diffsq = diff.*diff;
diffsum = sum(sum(diffsq));
% diffsum = sum(diffsum);
MSE = diffsum/(256*256);
% MSE = diffsum/128;
r = 65025/MSE;
```

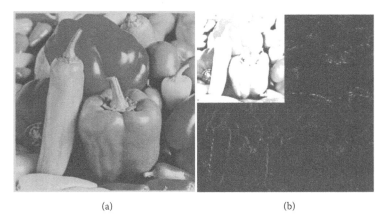

(a)   (b)

FIGURE 4.11
(a) Original peppers image. (b) Decomposed image at the first level.

```
PSNR = 10*log10(r);
sprintf('The MSE is%f',MSE)
sprintf('The PSNR is%f dB',PSNR)
[psnr1,mse1,maxerr1,L2rat1] = measerr(x,y);
```

The above program computes the peak signal-to-noise ratio (PSNR) with and without direct function. Readers are encouraged to try different thresholds and observe the quality and PSNR. You can obtain a threshold approximation, too, but because maximum energy is compacted in this, the resultant quality will be degraded.

The image obtained by decomposition is shown in Figure 4.11.

The resultant PSNR is 34.26 dB with the threshold values as shown in the program.

You will notice that blocking artifacts are removed in wavelet-based compression.

4.10 JPEG 2000 Image Compression Standard

JPEG 2000 (J2K) image compression gained popularity due to its high-quality, but slightly less compression ratio compared to JPEG. The features of JPEG 2000 are as follows:

- Superior low bit-rate performance
  - It provides excellent coding performance at bit rates lower than 0.25 bits per pixel for highly detailed gray-bits per level images.
  - At high bit rates, where artifacts become just imperceptible, JPEG 2000 has a compression advantage over JPEG of roughly 20% on average.

- Lossless and lossy compression
  - Both lossless and lossy compression are possible from a single compression architecture.
  - Lossless compression naturally is possible in the course of progressive decoding.

- Large images
  - It allows images greater than 64 × 64K without tiling.
  - It has a single decomposition architecture.
  - It encompasses the interchange between applications. There are no multiple modes for specific applications like JPEG.
  - Transmission is possible in noisy environments.
  - There is error robustness while designing the coding algorithm.

- Computer-generated imagery
  - The current JPEG is optimized for natural imagery and does not perform well on computer-generated imagery or computer graphics.

- Continuous-tone and bilevel compression
  - There is seamless compression of image components (e.g., R, G, or B), each from 1 to 16 bits.

- Progressive transmission by pixel accuracy and resolution
  - There is progressive transmission that allows images to be transmitted with increasing pixel accuracy or spatial resolution.
  - The image can be reconstructed with different resolutions and pixel accuracy as needed for different target devices such as in World Wide Web applications and image archiving.

- Real-time encoding and decoding

- Fixed rate, fixed size, and limited workspace memory
  - A fixed bit rate facilitates the decoder to run in real time through channels with limited bandwidth.

- The limited memory space is required by the hardware implementation of decoding.

The characteristics of J2K are as follows:

- License free
- Improved compression efficiency

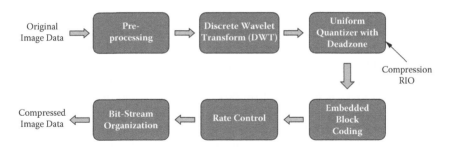

FIGURE 4.12
Block diagram of a JPEG 2000 encoder.

- Lossy and visually lossless compression
- Graceful degradation (soft blur in high-frequency areas)
- Scalability
- Dynamic bandwidth allocation
- Robust transmission
- Easy postproduction (easy proxy, easy editing)
- Region of interest
- Low latency
- Constant quality through multiple generations
- Same encoding and decoding power (rests on the same field-programmable gate array [FPGA] chip)

The encoding and decoding schemes used in JPEG 2000 are shown by the block diagrams in Figures 4.12 and 4.13, respectively.

Preprocessing consists of titling, which is optional, that divides an image into tiles. Next is the level offset in which the entire dynamic range is centered at zero. If done then this should also be added in the decoder. If images are in color, then they need to be converted in YCbCr, and the wavelet transformation is applied on every color space. After applying the wavelet

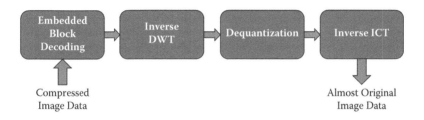

FIGURE 4.13
Block diagram of a JPEG 2000 decoder.

transformation, quantization is performed. They are quantized using a uniform quantizer with deadzone. Deadzone means that the quantization range about 0 is $2\Delta b$, resulting in more zeros. The embedded block coding uses embedded block coding with optimized truncation (EBCOT). Prior to coding, the subbands of each tile are additionally divided into comparatively small code-blocks (e.g., 64×64 or 32×32 samples), and every single code-block is encoded separately. The coding algorithm scans through the matrix of code-block quantization indices in a striped manner. The quantization indices of each code-block are not encoded at the symbol level but at the bit-plane level. A context-based adaptive binary arithmetic coder is used. Three coding passes are used: significance propagation, magnitude refinement, and cleanup. The resulting bit streams for each code-block are organized into quality layers. Then we alter the code stream in the rate control block in order to achieve the intended bit rate. The use of rate control guarantees the required number of bytes utilized by the code stream (truncation) at the same time guaranteeing the best-quality image that is possible. Finally, the bit stream is organized into main header, tile stream, tile header, packets, packet header, and so forth.

In the decoder, the reverse process takes place.

Questions

1. Give the scaling function for the "HAAR wavelet." Show that it is orthogonal to its translates.

2. Explain multiresolution decomposition using wavelet transform.

3. With the help of a block diagram, explain how we can achieve two-level wavelet decomposition.

4. Discuss the effects of the following properties of a wavelet on image compression:

   - Filter length
   - Orthogonality

Bibliography

A. Bovik, *Handbook of Image and Video Processing,* 2nd ed., Elsevier, Amsterdam, The Netherlands, 2005.

R. C. Gonzalez and R. E. Woods, *Digital Image Processing,* 2nd ed., Pearson Asia, Singapore, 2005.

R. C. Gonzalez, R. E. Woods, and S. L. Eddins, *Digital Image Processing Using MATLAB,* Pearson Prentice Hall, Englewood Cliffs, NJ, 2004.

S. Mallat, *A Wavelet Tour of Signal Processing,* 3rd ed., Elsevier, Amsterdam, 2003.

M. Rabbani and R. Joshi, An overview of the JPEG 2000 still image compression standard. *Journal of Signal Processing: Image Communication,* 17: 3–48, 2002.

R. M. Rao and A. S. Bopardikar, *Wavelet Transforms: Introduction to Theory and Applications,* Addison-Wesley-Longman, Boston, MA, 1997.

A. Said and W. A. Pearlman, A new, fast, and efficient image codec based on set partitioning in hierarchical trees. *IEEE Transactions on Circuits and Systems for Video Technology,* 6(3): 243–250, 1996.

K. Sayood, *Introduction to Data Compression,* 3rd ed., Morgan Kaufmann, Burlington, MA, 2006.

J. M. Shapiro, Embedded image coding using zerotrees of wavelet coefficients. *IEEE Transactions on Signal Processing,* 41(12): 3445–3462, 1993.

5

Image Compression Using Vector Quantization

Vector quantization is an efficient technique used for compressing images. It is based on the Shannon rate distortion theory, which says that better compression is achieved if samples are coded using vectors instead of scalars. The finite vectors of pixels are stored in a memory called codebook, which is used for coding and decoding the images. The image to be compressed is divided into blocks and called input vectors and are compared with vectors in memory called *codevectors* for matches based on some distance criteria. If the codevector matches the input vector, an index or address of memory location is stored or transmitted. Because the address has less bits than the codevector, compression is achieved. The decoding or decompression is the inverse of encoding. The quality of the reconstructed images depends upon proper design of the codebook. The algorithms used for the design of vector quantizers (VQs) (encoder and decoder), such as the oldest and famous Linde-Buzo-Gray (LBG) algorithm, are discussed in detail. Various types of VQs such as mean-removed, gain-shape, multistep, and others are presented. The VQ designs using image transforms such as discrete cosine and wavelet transforms are illustrated. The use of artificial neural networks in VQ design is also discussed. The performance of all designed codebooks are compared.

5.1 Introduction

In previous chapters, we studied the discrete cosine transform (DCT) and wavelet-based image compression schemes in detail. The main part of all lossy compression techniques is the quantizer. Quantization represents fine-resolution data by a coarse approximation. Therefore, the difference between fine and coarse resolution is quantization error, which is loss of information or distortion. In all compression schemes studied so far, scalar quantization is used. Scalar quantization takes place on a single sample followed by coding. This process may not involve memory. For example, pulse code modulation (PCM) does not require memory, whereas predictive quantization does.

An important result of Shannon's rate-distortion theory says that a good compression ratio is achieved if samples are coded using vectors instead of scalars. Even input data are random, instead of encoding the sequence of samples individually, coding using vectors provides an efficient code.

Quantization is a technique in which large sequences of samples or pixels are mapped into very small sequence or code. This is suitable for efficient storage or communication over the channel. Thus, the aim of the data compression is to reduce the bit rate/bits per pixel in order to minimize the storage requirements or bandwidth of the network. Data compression using VQ may be applied to any one-, two-, or three-dimensional data.

This chapter deals with the theory of VQ and how it is applied for image compression. Algorithms used for the design of VQs—that is, encoders and decoders used in VQ—will be presented. The different types of VQs used for efficient image compression are explained and compared. Before image compression, VQ was used for speech compression and recognition.

5.2 Theory of Vector Quantization

In VQ, if the source is a one-dimensional signal, then it is grouped into L consecutive samples; for a two-dimensional signal, which is an image, then it is divided into $n \times n$ subimages or blocks of L pixels. Thus, as shown in Figure 5.1, the dimension of a vector is L. This vector or block is an input to VQ. The complete VQ system consists of an encoder and a decoder. This is merely a memory, which stores a vector whose size is the same as that of an input vector. These encoders/decoders make up the *codebook*, and the vectors stored are called *codevectors*. The code words are selected or designed to be representative of the input vectors or blocks in an image to be compressed. Each codevector is assigned a binary index that is an address of

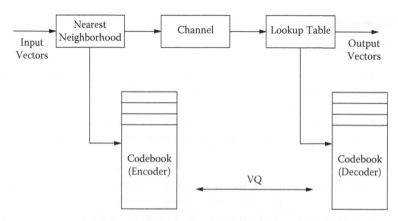

FIGURE 5.1
Block diagram of the vector quantizer process.

memory. An image block is compared with the codevectors. For the codevector that matches, an index or address is stored or transmitted. For matching, Euclidean or weighted mean square error (WMSE) distance is used. Thus, every subimage block is represented by an index. An encoded image thus contains only indices. Because an index contains a lesser number of bits, an encoded image is compressed.

The decoder is the same codebook used while encoding an image. The decoding or decompression is a simple process in which the codevector corresponding to every index is retrieved and an image is reconstructed. Since the codevectors are not the same as the original vectors, they provide a very limited representation of the input image. Thus, the decompressed image is degraded due to quantization error. The quality of a decompressed or decoded image depends solely on the design of the codebook and the number of codevectors.

For example, let an encoder and decoder be on 1024 locations and store the same number of codevectors. An index of location is 10 bits long. Every location contains 16 pixels of 8 bits each. The total number of bits will be 128. An image to be compressed is 256 × 256 pixel size. It is divided into blocks of 4 × 4 (16 pixels). The total number of blocks will be 4096. Every block after comparison with the codevector for the closest match will generate one index of 10 bits. Thus, the total number of bits required to store 4096 blocks will be 40,960, which is equal to 5 KB. The size of an uncompressed image is 64 KB, thus achieving a compression ratio of 12.8.

The compression ratio can be increased by reducing the size of the codebook or by increasing the size of the block, but the quality of the reconstructed image will be degraded. One can design a codebook of 512, 256, and even 64 size. However, the quality of the decompressed images will be deteriorated to a large extent. There is no reason to have a codebook size above 1024 as it will reduce the compression ratio.

Before we discuss more about VQ, let us define some terminology.

An encoder ε map's Q, k-dimensional space R^k to an index set I and finite subset of C of R^k; therefore,

$$Q: R^k \to C \quad C = (x_i^\wedge; i = 1,2,3\dots,N) \tag{5.1}$$

where C is the codebook and has N codevectors, also called *reproduction vectors*. The variable N also serves the size of the codebook.

The compression ratio or bit rate r linked with VQ depends upon N and vector size k. Thus, bit rate r is given as

$$r = \log_2 N / r \tag{5.2}$$

In the above example, bit rate r will be

$$r = \log_2 1024 / 16 = 0.625 \text{ bits per pixel (bpp)}$$

Thus, by using VQs, *fractional* bit rates can be obtained.

The closest codevector is selected by calculating squared error distortion and is given by

$$d(x,x^{\wedge})=\|x-x^{\wedge}\|^2 \sum_{i=0}^{N-1}(x-x^{\wedge})^2 \tag{5.3}$$

The Euclidean distance can also be used. The codevector that provides zero or minimum distortion is selected as the closest match. Apart from the Euclidean distance measure, sometimes weighted mean square error (WMSE) is used. The modified Itakura Saito distortion is used in voice coding applications (Gray 1984).

5.2.1 Advantages of Vector Quantization

Advantages of VQ include the following:

- VQ maps a group of inputs into a set of well-designed codevectors using minimum distance measures, whereas scalar quantization (SQ) maps each individual input to output with some minimum distortion.
- It is a very effective and efficient coding technique.
- It is secure. Only indices are stored and not actual data. For decoding one must have the same codebook.
- There is simplicity in its implementation, especially the decoder.
- It is useful in low-bit-rate applications.
- VQ can also be used as a pattern recognizer, where an input pattern is guessed by a predesigned set of a standard pattern.
- Fractional resolution can be achieved.
- A higher compression ratio is obtained.

5.2.2 Disadvantages of Vector Quantization

Disadvantages of VQ include the following:

- There is more encoding complexity because every vector has to be compared with a codevector for minimum distortion.
- The search for properly matching codevectors restricts the speed of encoding.
- There is a memory constraint.
- The reconstructed images may contain blockiness, which degrades the quality.

5.3 Design of Vector Quantizers

The performance of the compression scheme in terms of the quality of decompressed images is solely dependent upon the design of the codebook. The performance of VQs is normally measured by average distortion between input vectors and the selected codevectors. The smaller the average distortions, the better the performance. There are several techniques that have been developed during the last three decades for designing codebook. Many researchers are trying to bridge the gap between quality and compression ratio. The oldest and still most widely used algorithm for the design of the codebook was designed by Linde, Buzo, and Gray (Linde et al. 1980), popularly known as the LBG algorithm. Many VQs were designed by variants of LBG. In forthcoming sections, various types of codebook design techniques will be discussed in detail.

5.3.1 The Linde-Buzo-Gray Algorithm

As discussed in Section 5.2, the encoding process has a good rendering in the k-dimensional space. The set of codevectors defines a partition of R^k into N cells or partitions Vi, where $i = 1,2,3,...N$. If $Q(.)$ represents the encoding operator, then the ith partition is defined by

$$Vi = \{x \in R^k ; Q(x) = Ci\} \tag{5.4}$$

This type of partition that is unambiguously formed from the codebook with minimum distortion is called a *Voronoi partition* and is shown in Figure 5.2.

The encoding is the partitioning of space, and the big dots indicate the codevectors. The Linde-Buzo-Gray (LBG) algorithm is designed with two essential conditions:

1. For a fixed and same decoder codebook, an optimal encoder partition of R^k satisfies the minimum distance or neighborhood condition. Each input vector is mapped to the cell partition Vi having minimum distortion.

2. The *centroid* condition applies. For a given encoder partition cell, the optimum decoded code word is the centroid of that cell, where the centroid of the cell Vi is the vector C^* that minimizes $E\{d(x,y) \mid x \in Vi\}$.

$$C^* = \frac{1}{\|Vi\|} \sum_{l \in Vi} x_i \tag{5.5}$$

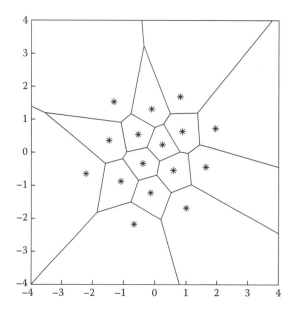

FIGURE 5.2
Voronoi diagram. (Courtesy of www.data-compression.com.)

The LBG design algorithm is a generalization of Lloyd's scalar quantization technique, which was used in a pulse code modulation design of a scalar random variable with a known probability density function and a square error distortion.

Thus, the LBG VQ algorithm is often called the *generalized Lloyd algorithm* (GLA). This algorithm was known earlier as the *k*-means algorithm by pattern recognition people.

The algorithm proceeds as follows:

- Start with a large training set ($\gg N$).
- Select N initial codevectors randomly. The size of the initial codebook is the same as the size of the final codebook.
- Encode all training vectors using the initial codebook (assigning subset of training vectors).
- For every codevector a cell of few input vectors will be associated. The centroid is computed for each cell.
- The centroids are used to generate an updated codebook.
- The procedure is repeated until convergence takes place. (One can terminate iterations when the codebook vectors are very near to local optimum.)

- For termination, check for the predetermined threshold of average distortion. If the calculated average distortion is greater than the threshold, iterate the procedure.
- The average distortion D is calculated as

$(D(l)-D(l-1)/D(l))$ Where l = iteration number D = average distortion

The initial codebook is a prime requirement of the LBG algorithm. As discussed, one can randomly select the vectors required to form the initial codebook. Various techniques are used to obtain an initial codebook. Linde et al. (1980) used the splitting technique to generate the initial codebook.

The splitting algorithm is as follows:

- Take the centroid of the entire training codevectors.
- Split the centroid (perturbation of the centroid by adding and subtracting perturbation vector $\in$).
- The encoding is done with these two centroids (subset is formed).
- The centroid is computed and replaced.
- Again centroids are split into two, and encoding is done.
- The procedure is repeated until a codebook of N codevectors is formed (symmetrically growing the codebook).

This algorithm provides a better codebook than designing an initial codebook using random codevectors. The codebook designed using the splitting technique can also be used directly.

5.3.2 Other Methods of Designing VQ

There are also other methods of designing an initial codebook. One such algorithm is the pairwise nearest neighbor (PNN):

- From the training set, compute the pairs of neighborhood codevectors and take the mean of the pairs. This will be a codevector. Thus, numbers of PNN codevectors equal to codebook size are considered.
- The codebook generated by PNN is used as the initial codebook in LGB.

The PNN algorithm has better computational efficiency. Codebooks designed using PNN can be used directly for encoding. There are a number of modifications suggested in LBG by many researchers.

Yogesh Dandawate and Madhuri Joshi have taken codevectors (blocks) from images that contain different features. From these blocks an initial codebook was prepared. Kohonen's self-organizing feature map neural network

(a)   (b)   (c)

FIGURE 5.3

(a) Original peppers image 256 × 256 pixel size. (b) Image compressed by codebook (1024 size, 4 × 4 vector size) designed using uniformly distributed random training vectors, PSNR = 14.83 dB, bpp = 0.625, CR = 12.5. (c) Image compressed by codebook (1024 size, 4 × 4 vector size) designed using training vectors from different images, PSNR = 26.33 dB, bpp = 0.625, CR = 12.5.

(Kohonen 1990) is also used for codebook design. As mentioned in a previous section, the major drawback of VQs is more search time and computational complexity during encoding. Some codebooks are designed to reduce search time.

Figure 5.3 shows an image compressed using the LBG designed codebook of 1024 size and 4 × 4 block size. The image compressed by the codebook designed using uniformly distributed random vectors is shown in Figure 5.3b. The quality is bad due to randomization of the training vectors. The data correlation is poor. Because images often contain correlated data, the training vectors are taken from different images that contain smooth variation of gray levels, edges, texture, various gray shades, and so forth, and the codebook is designed using LBG with an initial codebook. The quality of the decompressed image is acceptable and shown in Figure 5.3c.

Note that the MATLAB signal processing block set provides a VQ design block by invoking the dspquant2 command. The initial codebook is prepared by giving training vectors as input. The distortion measure, threshold, and iterations can be selected for better performance. The codebook generated is exported to the workspace as CB final. Using this codebook, encoding and decoding can be done. The program for VQ implementation is presented at the end of this chapter.

5.4 Tree-Structured Vector Quantizer

Buzo et al. introduced the concept of a tree-structured vector quantizer (TSVQ) for the purpose of fast search during the encoding process. The search time is reduced by a hierarchical arrangement of vectors. The concept

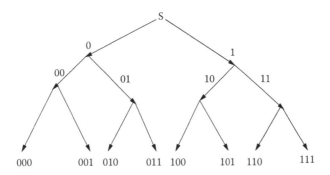

FIGURE 5.4
Binary search in tree-structured vector quantizer.

of a binary tree is used for the search. Codevectors are arranged according to the distance criteria in hierarchical order. The structure TSVQ is shown in Figure 5.4.

For example, let us take a codebook with eight codevectors. Let input vector x be searched for one of eight codevectors. The search is started at the top point S. Based on the positive or negative distance, further search is made to match with the closest. Here, based on the distance at the first level, the search is restricted to four locations in the codebook. If a match occurs in the first comparison, then the search is stopped. In TSVQ, there are three decision levels, whereas in full search there will be a maximum of eight decision levels. Implemented this way, $\log_2 N$ computations are required, and thus search speed is increased.

5.5 Mean-Removed Vector Quantizer

This VQ design requires less memory, is less complex, and performs better in comparison with the VQs discussed earlier. It has been observed that a codebook may have a similar vector that differs only in its mean value. This is the motivation behind the design of the mean-removed VQ. The size of the codebook is reduced by extracting the mean from the vectors and coding the extracted mean separately. This technique is proposed by Baker and Gray (1982, 1983). As shown in Figure 5.5, the mean of the input vector is calculated, and scalar quantization is performed on it. Then the mean-removed vector is given as input to VQ, which is designed using mean-removed vectors. The output is thus the VQ indices and the corresponding mean values.

At the decoder, the codevectors are retrieved from the mean-removed VQ and the mean is added to the vectors to form the reconstructed vector, which

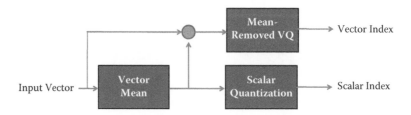

FIGURE 5.5
Mean-removed encoder.

is close to the input vector. The distortion can be less. This VQ is one type of structured VQ. Because mean-removed codevectors are less in comparison with the conventional codebook, the size of the codebook is less and this reduces the search time. This VQ uses a hybrid approach of quantization.

A variation of mean-removed VQ is mean-removed residual VQ. In this technique, the mean is scalar quantized and then subtracted from the input vector and applied to the codebook, which is designed using such quantized mean-removed training vectors. This VQ approach minimizes the blocking artifacts that are caused due to scalar quantization of the sample mean in the mean-removed VQ (MRVQ).

5.6 Gain-Shape Vector Quantization

Gain-shape VQ is similar to MRVQ. The gain term is extracted instead of the mean. The gain term is calculated as the Euclidean norm from input vector x and scalar quantized. The shape term is the normalization of input vector x by the gain term and VQ is performed on the shape. Output is thus an index of the codebook designed using shape training vectors. The gain and shape terms are calculated as

$$\text{Gain} = \| x \| \sqrt{\sum_{i=0}^{k} x[i]^2} \tag{5.6}$$

and

$$\text{Shape} = \frac{x}{\text{Gain}} \tag{5.7}$$

The mean-removed and gain shape can be combined to further reduce the size of the codebook and achieve an efficient compression ratio. The visual quality of images is better if compressed with mean-removed VQ than simple VQ and gain-shaped VQ when coded at 0.25 bpp with a 4 × 4 block size (Bovik 2005).

5.7 Classified Vector Quantizer

As discussed in Section 5.2.1, the major drawback of VQ-based compression is blockiness in decompressed images. Edge information is lost or degraded in small-size codebooks. In order to preserve the edges, it is necessary to have such codevectors present in the codebook. Gersho and Ramamurthi (1984) worked on this problem by classifying each image block vector into different classes or categories. Some of the classes are as follows:

1. Shade with smooth variation of gray levels, no significant gradient
2. Midrange class with moderate gradient without edge
3. Horizontal edge, in which vector blocks represent horizontal edges at different positions in the block such as bottom, mid, top, and so on
4. Diagonal edges with 45 degrees in which different blocks will present edges at different positions in the block like horizontal edges
5. Mixed edges with no definite single edge, but having a significant gradient

Similarly, vertical edges and edges at different angles are included as different classes. For every class, a separate codebook is designed using the LBG algorithm. This approach is popularly called the *classified vector quantizer* (CVQ). The block diagram of CVQ is shown in Figure 5.6. The input image blocks are classified using a classifier that employs gradient edge operators such as Sobel and Prewitt. The strength of the edge in a particular orientation is used to distinguish among the classes. Then encoding is done with

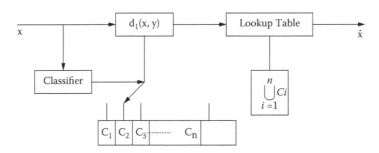

FIGURE 5.6
Classified vector quantizer.

a proper class VQ. Because only one subclass of the VQ is checked for every input vector, the computational complexity is reduced.

Gersho and Ramamurthi further extended the CVQ technique and used the discrete cosine transform (DCT) for every codevector in each class. The DCT coefficients are masked using zonal coding technique (Gonzalez and Woods 2005). The high-frequency coefficients are discarded, and the codevector size is reduced. Because most of the energy is packed by the low-frequency coefficients, especially for the blocks that belong to the midrange class, the distortion is less. Additional complexity involved is computation of DCT during encoding and inverse DCT during decoding.

Using CVQ, the quality of the decompressed images was observed to be good in comparison with VQ as discussed. The bit rate obtained was 0.6 to 1 bpp.

5.8 Multistage Vector Quantizer

In this approach, multiple codebooks are used for coding image blocks. The arrangement of the codebooks is multistage and is shown in Figure 5.7. The input vector is coded with the first codebook. The codebook is designed using the LBG algorithm. The difference between codevector and input vector, which is called the *residual vector,* is given as the input to the next stage and coded with codebook designed using training vectors that are residual codevectors. The residual vector of the second stage is coded in the third stage. The procedure can be carried out further if more stages are present. In the decoder, the codevector of the first stage is added with residuals in next stages, thus providing an image block of very little or no distortion. The quality of the decoded image is very good. The size of codebooks can be reduced, which in turn reduces search time. Due to multiple stages, an index generated may contain more bits. The compression ratio will be less in comparison with the VQs discussed so far. Using the appropriate codebook, the bit rate versus quality issue can be narrowed to some extent.

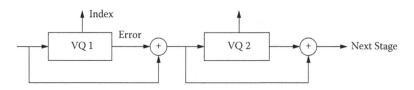

FIGURE 5.7
Multistage vector quantizer.

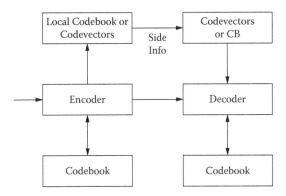

FIGURE 5.8
Adaptive vector quantizer.

5.9 Adaptive Vector Quantizer

If codebook size is large, the quality of the decompressed image is good. Having a super codebook will decrease the compression ratio or increase the bit rate. Instead, the size of the codebook is changed while encoding. Such a technique is called *adaptive vector quantization* and is shown in Figure 5.8.

The image is divided into nonoverlapping subimages or blocks, and for each subimage, a separate codebook consisting of about 16 to 64 vectors is formed. If the input vector does not match the codebook vector or the square error is well beyond the threshold, then it is coded with the locally generated representative vectors, and the codebook or codevector is transmitted as side information. This codebook is often called an *operational codebook*. This technique provides a better-quality image, but as mentioned above the bit rate lies between 1 and 1.25 bpp. The computational effort will also increase.

5.10 Hierarchical Vector Quantizer

All VQs including adaptive have the same size codevectors or blocks. According to Equation (5.2), the bit rate depends upon the size of the codebook and the size of the blocks. If block size is constant, only the size of the codebook will determine the bit rate. If the size of the block is also changed along with the codebook, then the bit can be reduced to a large extent. Hierarchical VQ design presents coding of the image with variable block size. In this coding technique, a quad-tree algorithm that is used in region splitting and merging-based segmentation is first used to partition the image

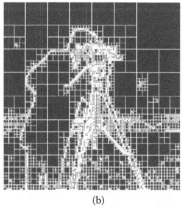

(a)   (b)

FIGURE 5.9
Quad-tree decomposition of cameraman image. (a) Original cameraman. (b) Quad-tree decomposed image into blocks of 2×2, 4×4, 8×8, and 16×16.

into blocks of size 2×2, 4×4, 8×8, 16×16, or more. The partitioning in formation of the image is represented by a quad-tree and is the side information, which can be transmitted. The typical quad-tree representation is shown in Figure 5.9. In the coding process, the small blocks, 2×2 and 4×4, are coded by using a typical VQ designed using LBG or any other algorithm. The larger blocks representing constant patterns are encoded by using VQ designed in the DCT domain (Section 5.7). Many high-frequency coefficients can be discarded, and thus the effective dimension of the blocks is reduced for fast computations. This technique was proposed by Nasarabadi (1985) and is also known as adaptive hierarchical VQ (AHVQ).

Note that the image in Figure 5.9 is decomposed using the qtdecomp command in the MATLAB Image Processing Toolbox. The information regarding blocks is obtained by using the qtgetblk and qtsetblk commands. Readers can try by using the help of qtdecomp.

AHVQ provides better quality at a reduced bit rate. The computational complexity is higher and suitable for compression of low- to midrange frequency images.

All the VQs we have discussed are structured VQs. They are memoryless vector quantizers of the scalar random variables.

5.11 Predictive Vector Quantizer

In images, normally the consecutive vectors are correlated with some relation. Better performance can be achieved if the correlation between the vectors is integrated in the encoder. Cuperman and Gersho (1982) have introduced a coding system involving a vector generalization of differential pulse code modulation. An encoder contents include a predictor and an error quantizer of vectors. An error vector is calculated by subtracting the predicted vector generated from the previous vector. The error vector is VQ encoded, which is designed using error training vectors. To enhance the performance, adaptive systems using classified VQ can be designed.

5.12 Transform Vector Quantizer

The aim of coding using transforms is to convert correlated image blocks into de-correlated transform coefficient blocks. In the previous chapter we studied JPEG image compression, which uses the DCT for transforming pixel blocks into the frequency domain. An image to be compressed is divided into 8 × 8 blocks, and the DCT is applied to every block. The maximum energy is compacted in the first few DCT coefficients, and high-frequency coefficients have negligible energy. Then, the block is quantized using scalar quantization. The high-frequency coefficients become zero due to quantization, and compression is achieved. In order to achieve better quality, adaptive quantization is used, where different blocks are quantized by different quantizers to retain certain useful information.

A similar concept is used while designing a codebook in the transform domain. The training vectors are transformed using unitary transforms like DCT and other wavelets. The transformed vectors are quantized in order to retain low-frequency and discard high-frequency coefficients. Thus, the size of the codevector will be reduced, which reduces computational complexity. Different quantization coefficients are based on different training vectors so as to make the transform adaptive. The decision of quantization is based on the class of input vector. Input vectors are classified based on the spatial and spectral activity content. Zonal techniques are used to determine the class (Gonzalez and Woods 2002). The transform-based codebook is observed to be more efficient than the spatial domain VQ, because of the better distributions of coefficients than the pixels. Instead of applying DCT on the training vectors, one can design a codebook in a normal way and then apply DCT on the codevectors. This gives slightly better results. Figure 5.10 shows images that have been compressed using a codebook designed with training vectors from different images and DCT applied on the same codebook with seven high-frequency coefficients discarded using zonal mask. The same mask is applied to all, irrespective of the activity, which means the quantization is not adaptive. Due to the energy compaction property of the DCT and well-defined distribution of the AC coefficients, the edges are preserved in DCT. The normal VQ process is more prone toward edge loss, so to see the effectiveness of the DCT-based VQ, an edge-based measure called the *mean edge*

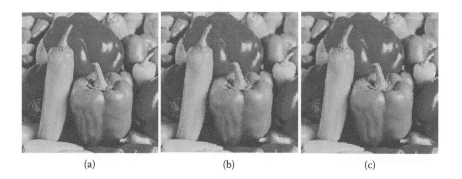

(a)   (b)   (c)

FIGURE 5.10
(a) Original pepper image size 256 × 256. (b) Compressed with codebook designed using LBG of 1024 size and 4 × 4 vector, PSNR = 26.33 dB, MESSIM = 0.55. (c) Compressed with DCT codebook of 1024 size and 4 × 4 vector and zonal mask, PSNR = 27.12 dB, MESSIM = 0.58.

structural similarity index (MESSIM) is used. The value ranges from 0 to 1. If close to 1, then all major edges in the decompressed image have been preserved. This was presented by Gaun Hao Chen et al. (2006) and uses the Sobel Operator. We have used the Canny edge operator and modified it for better performance. This edge quality measure is derived from the concept of the mean structural similarity index measure (MSSIM) developed by Z. Wang, A. Bovik et al. (2004), which takes the human visual system (HVS) into account for the determination of quality of decompressed images. This measure was developed because the metrics based on mean square error such as PSNR are not always correct indicators of the quality of decoded images.

As mentioned earlier, the number of discarded coefficients is fixed. If adaptive quantization is used, then the MESSIM might have been improved. The same codebooks are used to compress an image having high spatial and spectral activity. The results are shown in Figure 5.11.

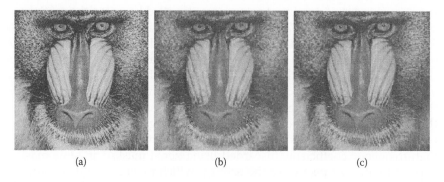

(a)   (b)   (c)

FIGURE 5.11
(a) Original mandrill image size 256 × 256. (b) Compressed with codebook designed using LBG of 1024 size and 4 × 4 vector, PSNR = 21.11 dB, MESSIM = 0.36. (c) Compressed with DCT codebook of 1024 size and 4 × 4 vector and zonal mask, PSNR = 22.60 dB, MESSIM = 0.50.

It has been observed that for high-frequency images, a DCT-based code-book works very well.

Wavelets are widely used in image compression. As discussed in Chapter 4, the image is decomposed into approximate and detail coefficients using low-pass and high-pass filters derived from scaling and wavelet functions, respectively. When an image is decomposed at the first level, then after subsampling it is divided into four regions: approximation, horizontal, vertical, and diagonal details (four subbands). Maximum energy is compacted in approximation coefficients. For image compression using VQ, four code-books corresponding to each subband are designed. The design can be done in the same way DCT codebooks are generated. The PSNR and MESSIM for the pepper image compressed using this VQ are 31.53 dB and 0.73, respectively. The major drawback is the compression ratio, which is now around 3 and bpp will be 2.4 due to four codebooks. To increase the compression ratio, detail coefficients are combined in one codebook and the approximation in another. The PSNR and MESSIM for the compressed peppers image using this VQ are 28 dB and 0.63, respectively. Still, wavelets perform much better than any other codebooks. Some of the coefficients can be thresholded and the size of the codevectors can be reduced further, which will increase bit rate.

Dandawate and Madhuri Joshi designed a codebook in which nearly similar codevectors are removed and residual vectors are inserted at the lower end of the codebook. The search will be first on the codevectors, the difference of codevector and input vector is calculated, and the search at the lower end of the codebook is performed. At decoder the codevector and the residue are added in order to get the image block close to the original. This will increase quality while keeping a low bit rate. Run-length coding can be applied on the indices to compress the data further.

5.13 Binary Vector Quantizer

Digital images of newspapers, weather maps, business documents, scanned literature, engineering drawings, and geographical maps normally use two-tone (black and white) values. Run length and entropy coding of the original data are often used for storage or transmission of such images. If the document image contains words in different fonts and paragraphs, then it is segmented into different sections. The image is further divided into blocks of N pixels, which is treated as a vector. Entropy coding such as Huffman coding is used in which shorter code words for frequently occurring vectors and longer for seldom occurring vectors are stored or transmitted. To reduce the bit rate, it is necessary to increase the number of pixels in a vector, but the complexity in Huffman coding will increase.

If some degradation in the quality is allowed, then for a given vector size, only a small number of possible pixel patterns of the subset of the block is allowed. Nonallowed patterns are replaced by the allowed subset pattern that matches it. This is known as a binary vector quantizer, and the codebook can be designed by using the LBG algorithm over a set of training sequences. The work is proposed by Knudson (1975). He carried out some experiments using newspaper texts using blocks of 8 × 8. It was found that a subset of 62 patterns was sufficient to obtain satisfactory image quality. A technique called block truncation coding (BTC) using a binary vector quantizer (BVQ) is used.

5.14 Variable-Rate Vector Quantization

The bit rate can be decreased by entropy coding of the indices. Huffman coding is often a choice in such conditions. The codevectors that occur frequently are assigned short-length indices and are rarely selected as assigned long-length indices. Making the index assignments in this way results in a low average bit rate, thus making coding more efficient. While using entropy coding, it is necessary to estimate the codevector probabilities $P(i)$. With these estimates, methods such as Huffman coding will assign to the ith index a code word whose length Li is approximately $-log, P(i)$ bits. Further optimization is entropy-constrained VQ (ECVQ) in which VQ and the entropy coder are designed together. Instead of finding the minimum distortion $d(x, x^\wedge)$ in the LBG iteration, one can find the minimum modified distortion:

$$Ji = d(x,x^\wedge) + Li \tag{5.8}$$

Employing this modified distortion, Ji, which is a Lagrangian cost function, effectively acts on Lagrangian minimization that seeks the minimum weighted cost of quantization error and bits. For further details readers are encouraged to consult works by F. Kossentini (1995).

5.15 Artificial Neural Network Approaches to Vector Quantizer Design

Before we discuss how neural networks are used to design VQ for better performance, let us briefly discuss artificial neural networks (ANNs).

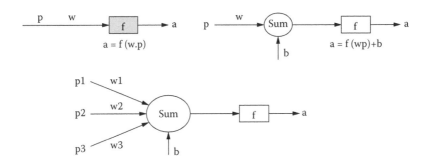

FIGURE 5.12
Neuron model.

5.15.1 Introduction to Artificial Neural Networks

Artificial neural networks are massive parallel computing networks with a large number of simple processing units, called neurons. A neuron is as shown in Figure 5.12. The input to any neuron can be a vector p. The neural unit is also associated with a weight vector w, and an activation function $f(w,p)$, which is the output of the neuron. Thus, its input vector, weight vector, and the activation function determine an overall input-output function of any neural network. In order to perform a particular task, neural networks are trained, and the process in which the magnitudes of weight vectors associated with the neurons are changed. The weights are modified based on the error derived from the training data. The techniques by which the weights are modified are categorized into *supervised* or *unsupervised*.

Artificial neural networks (ANNs) are computational frameworks or simple digital electronic models inspired by biological neuron networks in the human brain. ANNs will never replace biological neurons. Most ANNs use sets of elements connected by weighted links, thus forming a brain-like model. In the operation of neurons, the inputs are presented as weighted sums and passed through the nonlinear limiting function. This is also called the *squeezing function*. The ANNs are formed as interconnected layers consisting of input, hidden, and output layers. The initial weight values of all interconnections and inputs are normally initialized to random or zero.

Most of the ANNs operate in two stages. Various training or learning rules are used in order to classify the networks as supervised, unsupervised, or reinforced. During training, the weights of each layer in the network are adjusted for minimum mean squared error between target and actual output in case of supervised learning or to maximize differences between the output classes in case of unsupervised or competitive learning, which approximates the input pattern. Reinforced learning is a variant of supervised learning, in which the network is provided with only a cue on the network outputs and not actual outputs. In testing, the designed ANN generates outputs for the new inputs. We deal with unsupervised neural networks for the generation of the codebook.

Competitive learning algorithms are unsupervised training algorithms and can be used in VQ designs. The application of competitive learning algorithms, which is Kohonen's self-organizing feature map (KSOFM), is the basis of VQ design.

5.15.2 Competitive Learning Algorithm

In these algorithms, the weight vectors (or code words) w, associated with the neurons equal to the size of codevectors, are initialized with small random or midpoint values that are in range of pixel values, and the algorithm iterates through the input data, which serve as training data for a number of times, adjusting w, for each input vector x presented. The simple competitive learning (CL) algorithm can be described as follows:

1. Apply an input vector x.
2. Find the distortion $D_i = d(x,W_i)$ for all output neural units.
3. Select the output unit with the smallest distortion and label it as the winner and its weight vector as w_i^*.
4. Adjust the selected weight vector:

$$w_i^*(n+1) = w_i^*(n) + \varepsilon(n)[x(n) - w_i^*(n)] \tag{5.9}$$

where n is the training time index.

5. Repeat Steps (1) through (4) for all training vectors.

Note that the value selected for $\varepsilon(n)$ does not depend upon the magnitude of the data. The training rule moves the weight toward the training vector by some fractional amount, $\varepsilon(n)$. Typically, $0 < \varepsilon(n) < 1$ and decreases as training progresses.

One of the limitations of competitive networks is that some of the neurons may not always get allocated. Some neuron weight vectors may start from far and never win the competition, even though the training is continued for a long time. These neurons are the *dead neurons*, and they will not be useful. Biases are often used to overcome this problem. A positive bias makes a distant neuron more likely to win.

5.15.3 Kohonen's Self-Organizing Feature Maps

The Kohonen self-organizing feature map (SOFM) is a type of CL network that was proposed by Kohonen (1990). This maps the process according to topological feature maps that are formed in the brain. In the KSOFM or SOFM network, each neuron has an associated topological neighborhood of other neurons. During the training process, both the winning neurons as well as the neighborhood neurons of the winners are updated. The size of the neighborhood is decreased as training progresses until each neighborhood has only one unit.

The neighborhood of a unit in SOFM is a nothing set of connected neurons. For example, the connections may form grids of rectangular, hexagonal sizes. After training a winning unit is selected. The weight vector of each unit in the winning unit's neighborhood is then updated as follows:

$$w_i(n+1) = w_i(n) + \varepsilon(n,D)[x(n) - w(n)] \tag{5.10}$$

This uses a training rule that CL uses. The learning rate is a function of distance D from the selected unit and training time n. Generally, ε (n,D) decreases as the distance increases. The size of the neighborhood decreases as training progresses.

In the SOFM training algorithm, D_{max} is chosen to be large enough that all neurons are in the same neighborhood. Kohonen's self-organizing maps algorithm is as shown below:

1. Apply an input vector x.
2. Find the distortion $D_i = d(x,w_i)$ for all output units.
3. Select the output unit i^* with the smallest distortion.
4. Adjust the selected weight vector and its neighborhood of units:

$$w_i{}^*(n+1) = w_i{}^*(n) + \varepsilon(n,0)[x(n) - w_i{}^*(n)]$$

For all units that are less than D_{max} away from i:

$$w(n+1) = w_i(n) + \varepsilon(n,D)[x(n) - w(n)] \tag{5.11}$$

where n is the training time and $0 < \varepsilon$ $(n, D) < 1$.

5. Periodically decrease the extent of the neighborhood by decreasing D_{max} until $D_{max} = 0$.
6. Repeat Steps (1) through (5) for all training vectors.

Readers are referred to Krishnamurthy's paper (1990) for the example on SOFM. Also watching the demo from the MATLAB neural network can provide an understanding of how SOFM works.

The VQ is designed by taking vectors equal to the size of the codebook and applying the SOFM algorithm. The weight matrix generated represents the codebook. The efficiency of the codebook depends on chosen input vectors for training.

It has been noticed that the PSNR or the images compressed using VQ designed by SOFM is raised by 0.5 to 0.75 dB. By further refinement it outperforms other codebooks designed using LBG, except codebooks designed using transforms.

Dandawate and Joshi developed transform and wavelet-based codebooks using SOFM with image samples taken from different images as discussed earlier. As discussed in Section 5.12, a residual-based codebook is used in the transform domain to get better-quality images.

5.16 Concluding Remarks

After studying all VQ designs and their performances, it has been observed that transform coded and subband (wavelets) VQ work better than all other VQs. The quality and compression ratio gap can be narrowed if optimal codevectors are selected. The selection of these codevectors, developing fast search algorithms, and applying advanced techniques for variable bit rate coding are still active areas of research. The combined use of wavelets and curvelets will improve the performance of VQ for inputs not having classical shape edges. Genetic algorithms can be used in the selection of the best codevector, in which an initial population code word is created, evaluated, and then evolved through multiple generations using selection, crossover, and mutation operators in the search for a good solution for the code word. There are few papers that use contourlet transforms for the design. Ultimately, after developing an optimized codebook, the coders must be easily implemented in a field-programmable gate array (FPGA) or complex programmable logic device (CPLD) with much less component count and higher speed. The compression of color images can be performed by designing VQs in different color spaces (Dandawate et al. 2007) in spatial as well as in transform domains.

5.17 MATLAB Programs

The codebook generated using dspquant is stored in the workspace as finalCB.mat.

1. Program for encoding and decoding images using the LBG code-book designed using the VQ block in the DSP Blockset in MATLAB. The images taken are 256 × 256 gray images and divided into blocks of 4 × 4. However, the program can be generalized for any image size. The codebook size is 1 K, which is 1024 locations. Here 4096, 4 × 4 image blocks are used for training. They are obtained from the image (trainimage.tiff). One can take random training vectors using the rand or randi command. This command can be directly written in the training input box in the VQ design block.

```
clear all;
A = imread('trainigimage.tif');    % Read training image
A = rgb2gray(A);                   % convert in gray
[m,n] = size(A);
q = m/4;r = n/4;                   % Divide image into blocks
                                   of 4x4

p = 1;r1 = 1;
j1 = 1;j2 = 4;k1 = 1;k2 = 4;
while r1< = r
  while p< = q
  c(:,:,p) = A(j1:j2,k1:k2);
  p = p+1;
  k1 = k1+4;
  k2 = k2+4;
  end
  q = q+64;
  j1 = j1+4;
  j2 = j2+4;
  k1 = 1;k2 = 4;
  r1 = r1+1;
end
%%%%%%%%%%
for i = 1:4096                     % to make the count starting
                                   from 1 to 4096
  a(:,:,i) = double(c(:,:,i));
  g = a(:,:,i);
  rr = reshape(g,1,16);  % convert 4x4 blocks into 16x4096
                          matrix for the training purpose
  X(:,i) = rr;
end
%%%%%%%%%% Give X as training vectors to LBG based VQ in
dspquant2. Get
% finalCB as codebook in workspace
for i = 1:1024                     % Size of the codebook 1024
  a = round(finalCB(:,i));
  a = uint8(a);
  e = reshape(a,4,4);      % Codebook is designed for 4x4, one
                           can have 1x16 size
%%%%%%%%%%%%codevecotr also
```

```
  CB(:,:,i) = double(e);
end
%%%%%%%%%%   COMPRESSION AND DECOMPRESSION
I = imread('peppers.tif'); % Read image to be compressed
I = rgb2gray(I);
figure;
imshow(I);
[m,n] = size(I);      % Divide into blocks of 4x4. Other MATLAB
                      functions can also be
                      %%% used for dividing image into blocks
q = m/4;r = n/4;
p = 1;r1 = 1;
j1 = 1;j2 = 4;k1 = 1;k2 = 4;
while r1< = r
  while p< = q
  d(:,:,p) = I(j1:j2,k1:k2);
  p = p+1;
  k1 = k1+4;
  k2 = k2+4;
  end
  q = q+64;
  j1 = j1+4;
  j2 = j2+4;
  k1 = 1;k2 = 4;
  r1 = r1+1;
end
for i = 1:4096       % 4096 blocks from 256 x256 image
  b(:,:,i) = double(d(:,:,i));
end
%% program for Euclidean distance measurement (Searching for
proper codevector
z = 1;
for u = 1:4096
  for v = 1:1024
      x = b(:,:,u);
      y = CB(:,:,v);
      diff = x-y;
      diffsq = diff.*diff;
      diffsum = sum(sum(diffsq));
      dist = sqrt(diffsum);
      dis(v) = dist;
end
      distan = min(dis);
      for vv = 1:1024
        if (distan = =dis(vv))
             v1 = vv;
        end
    end
  CB1(:,:,z) = CB(:,:,v1);
  z = z+1;
```

```
end
% %% Program for reconstruction of the image (Decompressed
Image)
j11 = 1;j22 = 4;p1 = 1;k11 = 1;k22 = 4;
q1 = 1;q12 = 64;
while q1< = 64
   while p1< = q12
   B(j11:j22,k11:k22) = CB1(:,:,p1);
   p1 = p1+1;
   k11 = k11+4;
   k22 = k22+4;
   end
   q12 = q12+64;
   j11 = j11+4;
   j22 = j22+4;
   k11 = 1;k22 = 4;
   q1 = q1+1;
end
B = uint8(B);
figure;
imshow(B)              %% Display reconstructed image.
%% Calculate MSE and PSNR for quality Assessment.
x = I;
y = B;
diff = double(x)-double(y);      % difference of two images
diffsq = diff.*diff;
diffsum = sum(sum(diffsq));
% diffsum = sum(diffsum);
MSE = diffsum/(256*256);
% MSE = diffsum/128;
r = 65025/MSE;
PSNR = 10*log10(r);
sprintf('The MSE is%f',MSE)
sprintf('The PSNR is%f dB',PSNR)
```

2. Next is a program for designing the codebook using SOFM. The codebook size is 1024 and the codevector will be 4 × 4. The training vectors will be taken from the training image.

```
clear all;
A = imread('trainingimage.tif');
A = rgb2gray(A);
[m,n] = size(A);
q = m/4; r = n/4; % Divide image into blocks of 4x4, which
                    will be trained by SOFM
p = 1;r1 = 1;
j1 = 1;j2 = 4;k1 = 1;k2 = 4;
while r1< = r
```

```
    while p< = q
    c(:,:,p) = A(j1:j2,k1:k2);
    p = p+1;
    k1 = k1+4;                              \
    k2 = k2+4;
    end
    q = q+64;
    j1 = j1+4;
    j2 = j2+4;
    k1 = 1;k2 = 4;
    r1 = r1+1;
end
for i = 1:4096                              % total 4096 blocks
    a(:,:,i) = double(c(:,:,i));
end
for i = 1:4:4096      % Select one vector after every 4 vectors,
                      total 1024 are taken for
%%%%%%%training
net = newsom([0 255 0 255 0255 0255;0 255 0 255 0255 0255; 0
255 0 255 0255 0255; 0 255 0 255 0255 0255],[4]); %% Formation
of SOFM neural network, 0 255 is range of the%%% input pixel
values for 4 rows
    net.trainParam.epochs = 100;
    net = train(net, a(:,:,i)); %% Start training
    CB(:,:,in) = net.iw{1,1};        %% Weights acts as
    codevectors, transfer into codebook
    sprintf('block =%d', i) %%% Display Block number
end
```

Once the codebook is ready, the compression and decompression process is the same as explained in the previous program. Readers can
copy and paste the COMPRESSION AND DECOMPRESSION section from the previous program.

3. Next is the program for designing and testing the image in the discrete cosine transform (DCT) domain. From program 1, take the
training samples from cell c.

```
for i = 1:4096
h(:,:,i) = double(c(:,:,i));
t(:,:,i) = dct2(h(:,:,i)); % Apply DCT on the training Samples.
end
    for i = 1:4096                          %
    a(:,:,i) = double(t(:,:,i));
    g = a(:,:,i);
    rr = reshape(g,1,16);
    X(:,i) = rr;
end
```

```
%%%%%%%%%% Give X as training vectors to LBG based VQ in
dspquant2. Get
%%%%%%%%%% finalCB as codebook in workspace of size 1024.
for i = 1:1024
  a = round(finalCB(:,i));
  e = reshape(a,4,4);
  CB(:,:,i) = double(e);
end
%%%%%%%%%%COMPRESSION AND DECOMPRESSION
I = imread('peppers..tif');
I = rgb2gray(I);
figure;
imshow(I);
[m,n] = size(I);
q = m/4;r = n/4;
p = 1;r1 = 1;
j1 = 1;j2 = 4;k1 = 1;k2 = 4;
while r1< = r
  while p< = q
  d(:,:,p) = I(j1:j2,k1:k2);
  p = p+1;
  k1 = k1+4;
  k2 = k2+4;
  end
  q = q+64;
  j1 = j1+4;
  j2 = j2+4;
  k1 = 1;k2 = 4;
  r1 = r1+1;
end
for i = 1:4096
  b(:,:,i) = dct2(double(d(:,:,i))); % Applying DCT on testing
  image Samples
end
%program for euclidean distance measurement
%%%%%%%%%%%%%%%%%
z = 1;
for u = 1:4096
  for v = 1:1024
      x = b(:,:,u);
      y = CB(:,:,v);
      diff = x-y;
      diffsq = diff.*diff;
      diffsum = sum(sum(diffsq));
      dist = sqrt(diffsum);
      dis(v) = dist;
end
      distan = min(dis);
      for vv = 1:1024
    if (distan = =dis(vv))
```

```
    v1 = vv;
   end
 end
 CB1(:,:,z) = idct2(CB(:,:,v1)); %% Inverse DCT Program
 z = z+1;
end
% program for reconstruction
j11 = 1;j22 = 4;p1 = 1;k11 = 1;k22 = 4;
q1 = 1;q12 = 64;
while q1< = 64
  while p1< = q12
  B(j11:j22,k11:k22) = CB1(:,:,p1);
  p1 = p1+1;
  k11 = k11+4;
  k22 = k22+4;
  end
  q12 = q12+64;
  j11 = j11+4;
  j22 = j22+4;
  k11 = 1;k22 = 4;
  q1 = q1+1;
end
B = uint8(B);
figure;
imshow(B)
x = I;
y = B;
diff = double(x)-double(y);        % difference of two images
diffsq = diff.*diff;
diffsum = sum(sum(diffsq));
% diffsum = sum(diffsum);
MSE = diffsum/(256*256);
% MSE = diffsum/128;
r = 65025/MSE;
PSNR = 10*log10(r);
sprintf('The MSE is%f',MSE)
sprintf('The PSNR is%f dB',PSNR)
```

References

R. L. Baker and R. M. Gray, Differential vector quantization of achromatic imagery, presented at *Proc. Int. Picture Coding Symp.*, March 1983.

R. L. Baker and R. M. Gray, Image compression using non-adaptive spatial vector quantization, in *Proc. Conf. Rec. Sixteenth Asilomar Conf. Circuits, Syst., Comput.*, October 1982, pp. 55–61.

A. Bovik, *Handbook of Image and Video Processing*, 2nd ed., Elsevier Academic Press, New York, 2005.

G. -H. Chen, C. -L. Yang, L. -M. Po, and S. -L. Xie, Edge-based structural similarity for image quality assessment, *IEEE International Conference on Acoustics, Speech and Signal Processing, 2006. ICASSP 2006 Proceedings, South China Univ. of Technology, Guangzhou, China*, Vol. 2, pp. 933–936, IEEE, 2006.

Y. H. Dandawate, M. A. Joshi, and A. V. Chitre, Colour image compression using enhanced vector quantizer designed with self-organizing feature map, *Proceedings* of *International Conference on Information Processing*, ICIP-07, Bangalore, India, August 2007, pp. 80–85.

R. M. Gray, Vector quantization. *IEEE ASSP Magazine*, 1: 4–29, 1984.

D. R. Knudson, Digital encoding of newspaper graphics. *Electron. Syst. Lab., Mass. Inst. Technol., Rep. ESL-616*, August 1975.

T. Kohonen, The self-organizing maps. Invited paper, *Proceedings of IEEE 1990*, 78(9): 1464–1480.

Y. Linde, A. Buzo, and R. M. Gray, An algorithm for vector quantizer design. *IEEE Transactions on Communications*, 28(1): 84–95, January 1980.

Z. Wang, A. Conrad Bovik, H. Rahim Sheikh, and E. P. Simoncelli, Image quality assessment: From error visibility to structural similarity. *IEEE Transactions on Image Processing*, 13(4): April 2004.

Bibliography

V. Cuperman and A. Gersho, Vector predictive coding of speech at 16 kb/s. *IEEE Trans. Commun.*, COM-33, 1982, pp. 685–696.

Y. H. Dandawate, T. R. Jadhav, A. V. Chitre, and M. A. Joshi, Neuro-wavelet based vector quantizer design for image compression. *Indian Journal of Science and Technology*, 2(10): 56–61, October 2009.

Y. H. Dandawate and M. A. Joshi, Image compression using generic vector quantizer designed with Kohonen artificial neural networks: The quality improvement perspective. *International Journal of Information Processing*, 3(3): 45–54, 2009.

Y. H. Dandawate, M. A. Joshi, and S. M. Umrani, Performance comparison of color image compression based on enhanced vector quantizer designed using different color spaces. *Proceedings of International Conference on Intelligent and Advanced Systems*, Kualalumpur, Malaysia, November 2007, pp. 626–630.

Y. H. Dandawate and S. N. Londhe, Image compression using generic vector quantizer designed using transform coding: The quality analysis perspective. *Proceedings of Second International Conference on Computer and Electrical Engineering (ICCEE 2009)*, Vol. 2, Dubai, UAE, December 2009, pp. 624–628.

Demuth and Beale, *Neural Network Toolbox for MATLAB*, Version 4, *The MathWorks Inc.*, MA, USA.

A. Gersho and R. M. Gray, *Vector Quantization and Signal Compression*, Kluwer, Norwell, MA, 1992.

A. Gersho and B. Ramamurthi, Image coding using vector quantization. *Proceedings of IEEE International Conference on Acoustics, Speech, Signal Processing,* April 1982, pp. 428–431.

R. C. Gonzalez and R. E. Woods, *Digital Image Processing,* 2nd ed., Pearson Asia, Singapore, 2005.

R. C. Gonzalez, R. E. Woods, and S. L. Eddins, *Digital Image Processing Using MATLAB,* Pearson Prentice Hall, Upper Saddle River, NJ, 2004.

S. Hatmi, A modified method for codebook design with neural network in vector quantization based image compression. *Proceedings of the 2003 International Symposium on Circuits and Systems, ISCAS '03* (Vol. 2), 2003, pp. 612–615.

A. K. Jain, Artificial neural networks: A tutorial. *IEEE Computer Magazine,* March 1996, pp. 31–44.

M. A. Joshi and M. B. Khambete, Adaptive vector quantization based on quality criterion using Hosaka plot. *Proceedings of IEEE TENCON 1999,* pp. 754–757.

T. Kohonen, *Self-Organizing Maps,* 3rd extended ed., Springer, Berlin, 2001.

E. Kossentini, W. Chung, and M. Smith, Conditional entropy constrained residual VQ with application to image coding. *Transactions on Image Processing, Special Issue VQ,* February, 5: 311–321, 1996.

F. Kossentini, M. Smith, and C. Barnes, Necessary conditions for the optimality of variable rate residual vector quantizers. *IEEE Transactions on Information Theory,* 41(6): 1903–1914, 1995.

A. K. Krishnamurthy, S. C. Ahalt, D. E. Melton, and P. Chen, Neural networks for vector quantization of speech and images. *IEEE Journal on Selected Areas in Communications,* 8(8):1449–1457, October 1990.

N. M. Nasrabadi, Use of vector quantizers in image coding. In *Proceedings of IEEE International Conference on Acoustics, Speech, Signal Processing,* March 1985, pp. 125–128.

N. M. Nasrabadi and R. A. King, Image coding using vector quantization: A review. *IEEE Transactions on Communications,* 36(8): 957–971, August 1988.

B. Ramamurthi and A. Gersho, Image coding using segmented codebooks. *Proceedings of the International Picture Coding Symposium,* March 1983.

B. Ramamurthi and A. Gersho, Image vector quantization with a perceptually-based cell classifier. *IEEE Proceedings of the International Conference on Acoustic, Speech, Signal Processing,* March 1984.

K. Sayood, *Introduction to Data Compression,* 3rd ed., Morgan Kaufmann, Burlington, MA, 2006.

Y. Q. Shi and H. Sun, *Image and Video Compression for Multimedia Engineering, Fundamentals, Algorithms and Standards,* CRC Press, Boca Raton, FL, 2000.

6

Digital Video Compression

6.1 Introduction

Considerable research has been conducted in digital video coding technology over the last decade. Due to the growing availability of digital transmission links, progress in signal processing, very large-scale integration (VLSI) technology, and research in image/video compression, visual communications has become more viable than ever before. The driving force for the additional evolution of research in digital video coding is the wide range of applications such as video on demand, digital TV/HDTV broadcasting, and multimedia image/video database services.

Significant benefits achieved by video compression are that it offers an efficient and robust way to store or transmit the huge amount of data required to represent digital video. It also allows more effective use of transmission and storage resources.

The main goal of video coding is to reduce the average number of bits needed to represent a video sequence without affecting the video quality.

A significant amount of redundancy subsists between successive frames of video sequences. The redundancy that exists within frames is the spatial redundancy, and redundancy that exists between frames is the temporal redundancy. Video compression is achieved by exploiting these redundancies. The amount of redundancy contained in the image data and the actual compression techniques used for video coding greatly influence the performance of video compression techniques. Video compression can be categorized as lossless or lossy compression, which mostly depends on the application requirement. The objective of the lossless video coding is to reduce video data for storage and transmission while retaining the quality of the original video.

Medical imaging and satellite imaging are examples of lossless applications where transmission or storage of the highest-quality images is required.

As compared to lossless video coding, the goal of lossy video coding is to reduce the bit rate for storage and transmission. High video compression is achieved by degrading the video quality. The compression ratio depends on the target bit rate of the channel. High compression is the requirement of the smaller target bit rate of the channel.

Temporal and spatial redundancies exist within video sequences.

Interelement correlation is the basic statistical property of the image sequence including the assumption of simple correlated translatory motion between consecutive frames. The intraframe coding technique is one in which the magnitude of a particular image pixel can be predicted from a nearby pixel within the same frame, whereas in the interframe coding technique it is predicted from nearby frames. An exploration of a proper method to exploit the temporal redundancy completely changes the scenario between compression of still pictures and sequence of images. These methods are the foundation for very high performances in image sequence coding when compared to still image coding. Motion estimation and compensation techniques have shown their efficiency to reduce the temporal redundancy in this respect.

6.2 Digital Video Data

A video stream is continuous in both spatial and temporal domains. Spatial and temporal sampling of the video stream represents the digital video data. This sampled version of video data is shown in Figure 6.1. An image sampled in the spatial domain is typically represented on a rectangular grid. A temporally sampled video stream is represented as a series of still images sampled at a regular interval of time. These still images are called *frames*. The spatiotemporal sample value termed *pixel* is represented as a positive digital number that describes the brightness (luminance) in case of a monochromatic image and color components in case of a color image.

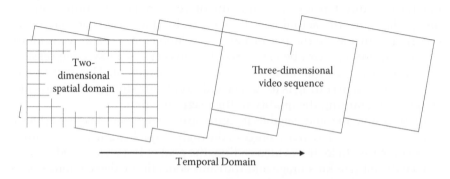

FIGURE 6.1
Three-dimensional video sequence.

6.3 Video Compression Techniques

Uncompressed video is composed of a huge amount of data. Due to limited communication and storage capabilities, handling uncompressed video data is expensive. For example, an HDTV video signal with 720 × 1280 pixels/ frame with progressive scanning at 30 frames/second required the following channel bandwidth to transmit the uncompressed data:

$$\left(\frac{720 \times 1280 \ pixels}{frame} \right) \left(\frac{30 \ frames}{sec} \right) \left(\frac{3 \ colors}{pixel} \right) \left(\frac{8 \ bits}{color} \right) = 6.6 \ Gb/s \quad (6.1)$$

Then, 20 Mb/s is the available channel bandwidth for HDTV. With the available bandwidth, if we want to transmit uncompressed HDTV data, compression is required by a factor of 35.

A digital versatile disk (DVD) can store only a few seconds of uncompressed video data. Without video and audio compression, it would not be possible to store video data on a DVD.

Favorably, digital video contains a significant amount of spatial and temporal redundancies. Compression is achieved by eliminating or reducing these redundancies. More specifically, video compression is achieved by exploiting four types of redundancies: perceptual, temporal, spatial, and statistical.

6.4 Perceptual Redundancies

Human eyes are not perfect sensors for discriminating small differences in an image. These differences can be discarded without affecting the image quality. Estimation and exploitation of these perceptual redundant data enhance the performance of the compression technique. Video signals are composed of spatial and temporal perceptual data that the human visual system would not be able to distinguish.

6.4.1 Temporal Perception

Persistence of vision states that the human eye always retains images for a fraction of a second (around 0.04 second). Human perception of motion is the result of persistence of vision, and this results in humans being able to recognize a sequence of frames as a continuous moving picture. If frames are displayed at a rate greater than the persistence of vision, the human visual system can recognize motion in a video sequence. Frame rate is decided by the content of the video sequence. It is necessary to display a video sequence with high motion content at a greater frame rate to avoid the flickering effect. The persistence property can be exploited by selecting the proper frame rate needed just to maintain motion in the video sequence. For some applications, to manage with the available bit rate of the channel, video is subsampled in the temporal direction.

6.4.2 Spatial Perception

The human eye is less sensitive to the higher spatial frequency components than the lower frequencies. In addition to this, the human eye is more sensitive to changes in brightness than to chromaticity changes. This property of the eye can be exploited to remove high-frequency components without affecting visual quality.

The properties of human visual perception thus allow exploitation of spatial and temporal perceptual redundancies.

6.5 Exploiting Spatial Redundancies (Intraframe Coding Technique)

Exploiting spatial redundancies achieves video compression by reducing high spatial correlation between neighboring pixels within a video frame. Such a technique is also known as the intraframe coding technique or still-image coding technique. Prediction based on neighboring pixels, entitled as intraprediction and transform-based coding, are some of the techniques used for intraframe coding.

6.5.1 Predictive Coding

In predictive coding, the redundancy in video data is determined from the correlation of neighboring pixels, which exists within frames or between frames. Predictive coding is advantageous if the correlation is strong enough among the spatially and temporally adjacent pixels. The basis of predictive coding is to predict a pixel value that needs to be coded from previous pixel values which were already transmitted. The prediction error (i.e., the difference between the actual pixel and the predicted pixel value) is quantized and then it is entropy coded. This is a very popular form of predictive coding known as differential pulse code modulation (DPCM). In view of the fact that neighboring pixels are highly correlated, the prediction error tends to be small. The prediction error is then combined with variable run-length coding (i.e., entropy encoding). Variable-length coding encodes the more common values with a shorter code and the less common values with a longer code. The coding efficiency of this system depends to a large degree on the prediction accuracy. The more accurate the prediction is, the smaller the prediction errors will be, and hence the fewer number of bits are required to represent them.

6.5.2 Transform Coding

Transform-based coding is the widely adopted method for image and video compression. This coding method achieves compression by exploring spatial correlation. De-correlation of image data is the objective of transform coding. These de-correlated transform coefficients are then encoded instead of the original pixels of the image. In this method, the input image is split into nonoverlapping blocks of pixels. These blocks of pixels are then transformed into a set of transform coefficients. An effective transform should compact the energy in the block of pixels into only a few of the corresponding coefficients. These transform coefficients are then coded for the transmission. When selecting a transform, three main properties need to be considered: good energy compaction, data-independent basis functions, and fast implementation. The performance of many suboptimal transforms with data-independent basis functions has been studied (Clarke 1985). Examples are the discrete Fourier transform (DFT), the discrete cosine transform (DCT), the Walsh-Hadamard transform (WHT), and the Haar transform. Among many possible alternatives, it has been demonstrated that the discrete cosine transform (DCT) is the most successful transform for still image and video coding due to its closest energy-compaction performance (Ahmed, Natrajan, and Rao 1984).

And hence most image and video coding standards adopted DCT-based implementations.

6.6 Exploiting Temporal Redundancies (Interframe Coding Technique)

Two successive frames of a video sequence are highly correlated as these are temporally sampled. Two such consecutive frames in a video are shown in Figure 6.2. Temporal sampling frequency and relative object motion within frames will decide the correlation factor of frames. A highly correlated video signal results in high temporal redundancy. Exploitation of these temporal redundancy gains high compression ratio in video coding, which is the principle for interframe video coding.

FIGURE 6.2
Successive frames in a video.

6.6.1 Interframe Predictive Coding

Owing to the fact that two successive frames of a video sequence are highly correlated, we can predict the current frame from the previously coded frame. This is called the *predictive video coding technique.* For efficient video coding, computation of prediction of a frame plays a significant role. Thus, in the interframe coding technique, frames are coded based on some relationship with other video frames (i.e., coding exploits the interdependencies of video frames). Interframe predictive coding is shown in Figure 6.3.

Frame differencing is the simplest form of predictive coding. As the two successive frames of a video sequence are highly correlated, if we take the difference between two consecutive frames, the resultant error signal is small, which needs to be encoded. This assumption works better if the object

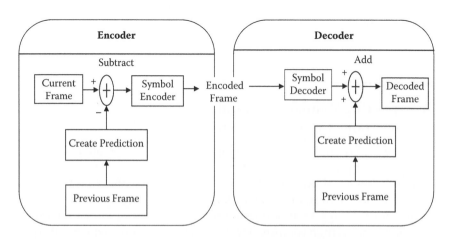

FIGURE 6.3
Interframe predictive coding.

motion is limited in consecutive frames. A simple frame difference gives a poor result even for a small motion of one to two pixels for the corresponding object in successive frames. To overcome this loss of motion information, motion compensation is widely adopted in video compression. The object motion in the real world is complex, but a simple translational motion is assumed for the purpose of video compression (Richardson 2003).

6.6.2 Motion-Compensated Prediction

To reduce temporal redundancies between frames, motion-compensated prediction is a powerful tool and is widely used in video coding standards. The concept of motion compensation (MC) is based on the estimation of motion between two video frames. If we consider small $N \times N$ pixel regions of an image, the extent of changes in this small region of two successive frames of a video is small. For an assumption that all elements in a video scene are approximately spatially displaced, the $N \times N$ region in a current frame can be better predicted from a previous frame by displacing the $N \times N$ region in the previous image by an amount representing the object motion. The amount of this displacement depends on relative motion between the two frames (Richardson 2003). The motion between frames can be described approximately by a limited number of motion parameters, which are called *motion vectors* (MVs). Motion estimation (ME) is the process in which motion of the current block is predicted from the previous frame block, and this is depicted in Figure 6.4. Generally, prediction error and motion vectors are transmitted to the receiver for the prediction of the current block. This process is the well-known motion-compensated (MC) prediction technique.

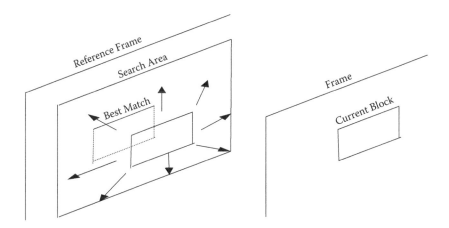

FIGURE 6.4
Motion estimation.

6.7 Exploiting Statistical Redundancies

The final process of video compression is to compress the quantized coefficients by an entropy coder, which removes statistical redundancy in the data. Owing to its simplicity and efficiency, variable length coding (VLC) is widely deployed for the entropy coding. In the VLC method, a shorter code word is assigned to values with higher probability. By this approach, compression efficiency is thus improved by exploiting statistical redundancies.

The simplest compression method is to remove the spatial redundancy that exists in each frame. The compressed frames are then sent out independently. This is called intraframe coding. However, the efficiency of intraframe-only coding is not very high because temporal redundancy between frames in a video is not removed. Temporal redundancy between frames is removed by interframe coding. When both intraframe and interframe coding are used, it is called *hybrid coding*. Hybrid coding achieves great coding efficiency. Popular video coding standards like ISO/IEC MPEG (ISO/IEC JTC 1 1993, 1994, 1999, 2000, 2004) and ITU-T H.26X (ITU-T 1990, 1993, 1995, 1998, 2000; ISO/IEC JTC 1 2003, 2004, 2005) series are hybrid coding schemes.

6.8 Hybrid Video Coding

Video compression algorithms adopted a combination of the intraframe and interframe techniques, which were discussed in the last sections in order to reduce redundancies. The hybrid video compression technique uses transform coding for exploiting spatial redundancies and motion compensation to exploit temporal redundancy (Sikora 1999). The basic components of a generalized hybrid video encoder are shown in Figure 6.5. The input video frame is split into disjoint blocks of pixels and the frame is encoded block by block. The first frame of the video sequence is encoded using the intraframe coding mode without motion compensation as no reference frame is available. Succeeding this, spatial domain pixels are converted into transform coefficients using the transformation module (T). This is followed by the quantization (Q) of transform coefficients with subsequent run-length coding and entropy coding (VLC). The previously coded frame $(N - 1)$ 1 is stored in a frame store (FS) at both the encoder and decoder module. The motion estimation (ME) module evaluates the motion vector for each block of pixels. Motion compensation (MC) is performed on subsequent frames to exploit temporal redundancies. The motion compensation (MC) module uses the motion vector and obtains a predicted block from the reference frame. Prediction error is then encoded.

Hybrid video coding is the technique accepted by most of the video compression standards today. Each standard specifies different tools for performing motion estimation, transform coding, quantization, and variable-length coding.

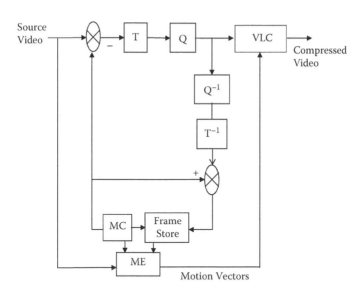

FIGURE 6.5
Hybrid video encoder.

6.9 Block Matching Motion Estimation

Motion estimation and compensation is the most critical process in the entire video coding development. Block-matching motion estimation (BMME) (Jain and Jain 1981) is the most typical method adopted for motion estimation. It has been proven that the block-based motion estimation technique is the best compromise between complexity and quality, and so it has been recognized by all the video coding standards proposed to date. The motive for block-based motion estimation is that it is easy to implement in hardware. Block-based motion compensation works on the following assumptions:

- The illumination is uniform along motion trajectories.
- The problems due to uncovered areas are neglected.

For the first assumption it neglects the problem of illumination change over time, which includes optical flow but does not correspond to any motion. The second assumption refers to the uncovered background problem. Basically, for the area of an uncovered background in the reference frame, no optical flow can be found in the reference frame. Although these assumptions do not always hold for all real-world video sequences, they are still used on the basis of many motion estimation techniques.

Figure 6.6 shows how the block matching motion estimation technique works. Each frame of size $M \times N$ is divided into square blocks $B(i,j)$ of size (b

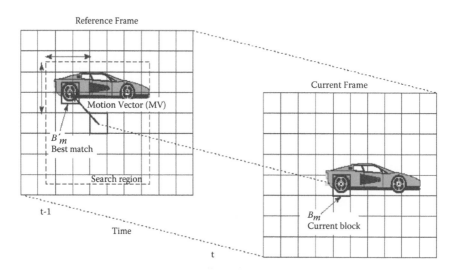

FIGURE 6.6
Block-matching motion estimation.

$\times b$) with $i = 1,\ldots, M/b$, and $j = 1,\ldots, N/b$. For each block B_m in the current frame, a search is performed on the reference frame to find a matching based on a block distortion measure (BDM). The motion vector (MV) is the displacement from the current block to the best matched block in the reference frame. Usually a search window is defined to confine the search. The same motion vector is assigned to all pixels within the block:

$$\forall \vec{r} \in B(i, j), \quad \vec{d}(\vec{r}) = \vec{d}(i, j) \tag{6.2}$$

where the image intensity at pixel location $\vec{r} = (d_x, d_y)^T$ and at time t is denoted by $I(\vec{r}, t)$ and $\vec{d} = (d_x, d_y)^T$ is the displacement during the time interval Δt.

Suppose a block has size $b \times b$ pixels and the maximum allowable displacement of a MV is $\pm w$ pixels in both horizontal and vertical directions. There are $(2w + 1)^2$ possible candidate blocks inside the search window. The basic principle of block-matching algorithm is shown in Figure 6.7.

A matching between the current block and one of the candidate blocks is referred to as a point being searched in the search window. If all the points in a search window are searched, the finding of a global minimum point is guaranteed.

The parameters of the BMA which has impact on performance and accuracy of motion estimation and compensation are distortion function, block size, and maximum permissible motion displacement, also specified as the search range.

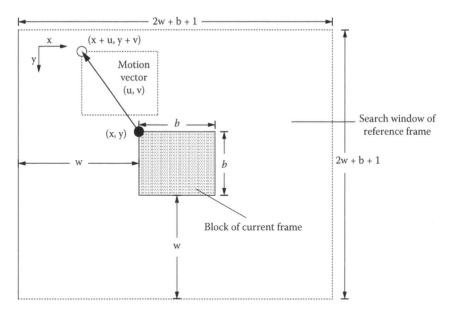

FIGURE 6.7
Block-matching method.

6.9.1 Block Distortion Measure

The success of the block-matching method mostly depends on substitution of the block in the compressed frame for the replacement of the original block. The block distortion measure function plays an important role in finding out the similarity between the target block and candidate blocks.

Assume that F_t is the current frame and F_{t-1} is the reference frame. $F(x,y)$ is the intensity of a pixel at (x,y) in frame F. Each candidate block is located at $(x+w_x, y+w_y)$ inside a search window of size $\pm w$ pixels such that $-w \le w_x, wy \le +w$. The optimum motion vector that minimizes the block distortion measure (BDM) function is (u,v). To estimate the better match, several criteria are defined. Extensively used criteria for block-based motion estimation are

- Mean square error (MSE)
- Sum of absolute difference (SAD)

The mean square error of a block of pixels computed at a displacement (w_x, w_y) in the reference frame is given by

$$MSE(w_x, w_y) = \frac{1}{N \times N} \sum_{i=x}^{x+N-1} \sum_{j=y}^{y+N-1} \left[F(i,j) - F_{t-1}(i+w_x, j+w_y) \right]^2 \qquad (6.3)$$

To find the displacement of the block in the reference frame, the MSE is calculated for each displacement position (w_x, w_y). Displacement position (w_x, w_y) is examined within a quantified search range in the reference frame (Gorpuni 2009). The position that provides the minimum value of MSE is identified as motion vector and is given by

$$(u,v) = \min_{-w \le w_x, w_y \le +w} MSE(w_x, w_y)$$

(6.4)

From Equation (6.3), it is clear that computations that are required to calculate MSE for each block for every search position are N^2 subtractions, N^2 multiplications, and ($N^2 - 1$) additions.

MSE is the Euclidian distance between current and reference blocks. It is considered to be better BDM because it gives results that are closer to our visual perception. The shortcoming of MSE is that it is more complex than other distortion measures as it requires square operations.

The sum of absolute difference (SAD) also yields positive error values similar to MSE. But instead of summing up the squared differences, the absolute differences are summed up. The SAD at displacement (w_x, w_y) is defined as

$$SAD(w_x, w_y) = \sum_{i=x}^{x+N-1} \sum_{j=y}^{y+N-1} \left| F(i,j) - F_{t-1}(i + w_x, i + w_y) \right|$$

(6.5)

The motion vector is defined as

$$(u,v) = \min_{-w \le w_x, w_y \le +w} SAD(w_x, w_y)$$

(6.6)

For every block at each search position the SAD criterion needs N^2 additions and N^2 subtractions (Gorpuni 2009). No multiplication is involved in calculating the SAD and for this reason the SAD criterion is preferred for hardware implementation.

6.9.2 Block Size

Block size is a significant parameter of the block-matching technique. Smaller block size succeeds for better prediction quality. Within a block of small size, there is a constraint on object motion moving in different directions, if the block size is small it gives better approximation with regard to small translation. The choice of block size has an impact on computational complexity of motion estimation. Larger block size results in fewer computations at the expense of poor prediction quality. Smaller block size results in large motion overhead to get better quality. Most video coding standards use a block size of 16 × 16 as a compromise between prediction quality and motion overhead.

6.9.3 Search Range

The search range is the maximum permissible motion displacement. The prediction quality and complexity of the block-matching technique depend on the search range.

For a fast-moving object where motion extends on larger areas, a small w results in poor compensation and accordingly poor prediction quality. With an increase in the computational complexity, the choice of large $\pm w$ results in better prediction quality (Gorpuni 2009). A larger w also results in more motion vectors and consequently a slight increase in motion overhead (Liu and Zaccarin 1995). As a compromise between quality and complexity, for low-bit-rate applications a maximum allowed displacement of $w = \pm 7$ pixels is sufficient.

6.10 Motion Search Algorithms

The search strategy used to find the best-matching predictor block, called a *motion search algorithm*, is not being specified by the video coding standard. Many motion search algorithms have been proposed over the past few decades. One possible motion search is an exhaustive (or full) search that searches all possible blocks in the search region; this strategy is a natural one and guarantees a global minimum BDM within the search region. Unfortunately, the exhaustive search is computationally very expensive. Due to its regularity, a full search is an attraction for hardware designers.

Fast-motion search algorithms are commonly used to reduce computation in the motion estimator, especially for software encoders where the regularity of the search pattern is usually not as great a concern as in hardware implementations. Fast searches require only a fraction of the computation of the exhaustive search, but they cannot guarantee a global BDM.

Here, we discussed four different fast-motion search algorithms. These four search algorithms are the two-dimensional logarithmic search, three-step search, the cross search, and one-at-a-time search. This section describes each of these motion-search algorithms in greater detail and presents performance results for these searches based on simulations for motion estimation.

6.11 Exhaustive Search

The exhaustive search compares each block of the given search region of the reference frame against the target block of the current frame. Although the algorithm is straightforward, the implementation of the exhaustive search is important. If, in the search region, two candidate blocks both yield the same BDM value with respect to the target block, the candidate block having the shortest motion vector is selected. Thus, the exhaustive search in the basic encoder is implemented in a spiral fashion, beginning at the "center" with the candidate block whose coordinates are the same as the target block, and going outward from there. In this fashion, candidate blocks with shorter motion vectors are evaluated first. The first candidate block with the lowest BDM value is the best-matching block.

6.12 Two-Dimensional Logarithmic Search Algorithm

A two-dimensional logarithmic (TDL) search was introduced by Jain and Jain in 1981. TDL assumes that the block-matching error value increases monotonically in all directions away from the true optimal block. The initial step size s is $\lceil w/4 \rceil$ (where $\lceil \ \rceil$ is the upper integer truncation function) and w is the search range in either direction. To find out the best match the block at the center of the search area and the four candidate blocks at a distance s from the center on the x and y axes are compared to the target block. The five positions form a pattern similar to the five points of a Greek cross (+). Thus, if the center of the search area is at position $[0, 0]$, then the candidate blocks at position $[0, 0]$, $[0, +s]$, $[0, -s]$, $[-s, 0]$, and $[+s, 0]$ are examined. Figure 6.8 shows the search pattern of the TDL search algorithm.

The decision to reduce the step size depends on the resultant previous minimum distortion point. If the minimum distortion measure point of the previous step is the center (cx, cy) or the current minimum point reaches the search window boundary, the step size is reduced by half. Otherwise, we need to continue with the same step size. When the step size is reduced to one, then all eight blocks around the center position $[cx, cy]$, which are $[cx - 1, cy - 1]$, $[cx - 1, cy]$, $[cx - 1, cy + 1]$, $[cx, cy - 1]$, $[cx, cy + 1]$, $[cx + 1, cy - 1]$, $[cx + 1, cy]$, and $[cx + 1, cy + 1]$, are examined, the minimum distortion measure point of these is determined to be the best match for the target block and thus the algorithm is terminated. Otherwise (step size greater than one) the candidate blocks at positions $[cx, cy]$, $[cx + s, cy]$, $[cx - s, cy]$, $[cx, cy + s]$, and $[cx, cy - s]$ are evaluated for distortion measure. An experimental result proves that the algorithm performs well in large motion sequences because search points are quite evenly distributed over the search window.

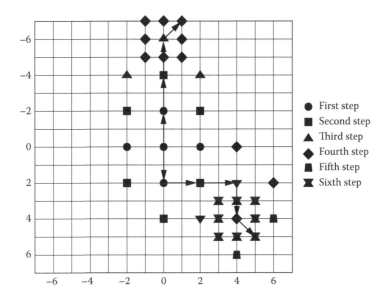

FIGURE 6.8
TDL search.

6.13 Three-Step Search Algorithm

This algorithm is based on a coarse-to-fine approach with logarithmic decreasing in step size as shown in Figure 6.9. The three-step search algorithm (TSS) tests eight points around the center (Koga and Ishiguro 1981).

For a step size equal to d and a center position $[cx, cy]$, nine positions $[cx - d, cy - d]$, $[cx - d, cy]$, $[cx - d, cy + d]$, $[cx, cy - d]$, $[cx, cy]$, $[cx, cy + d]$, $[cx + d, cy - d]$, $[cx + d, cy]$, and $[cx + d, cy + d]$ are evaluated for distortion measure. After each stage, the step size is halved and the minimum distortion of that stage is chosen as the starting center of the next stage. The procedure continues until the step size becomes one. In this manner, TSS reduces the number of searching points as equal to $[1 + \lceil \log_2 (d + 1) \rceil]$. One problem that occurs with the TSS is that it uses a uniformly allocated checking point pattern in the first step, which becomes inefficient for small motion estimation.

6.14 Cross-Search Algorithm

The cross-search algorithm (CSA) proposed by Ghanbari (1990) is a logarithmic step search algorithm using a saltire cross (X) searching pattern in each

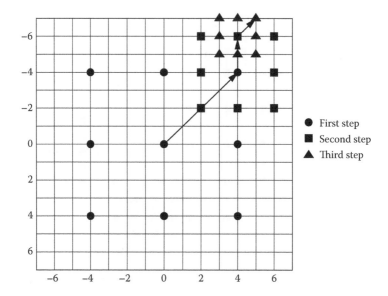

FIGURE 6.9
Three-step search algorithm.

step. The CSA is presented in Figure 6.10. The initial step size is half of maximum motion displacement w. The step size is halved at each stage until the step size turns out to be one. At the final stage, however, the endpoints of a Greek cross (+) are used to search areas around the top-right and bottom-left corners of the previous stage. And a saltire cross (x) is used to search areas around the top-left and bottom-right corners of the previous stage.

The CSA requires $[5 + 4 \lceil \log_2 w \rceil]$ comparisons where w is the largest allowed displacement. The algorithm has a low computational complexity. It is, however, not the best in terms of motion compensation.

6.15 One-at-a-Time Search Algorithm

The one-at-a-time search algorithm (OTA) is a simple but effective algorithm that has a horizontal and vertical stage (Srinivasan and Rao 1985). OTA starts its search at the search window center. The center points and its two horizontally adjacent points [i.e., (0, 0), (0, −1), and (0, +1)] are searched. If the smallest distortion is for the center points, start the vertical stage, otherwise look at the next point in the horizontal direction closer to the point with the smallest distortion, and continue in that direction until you find the point with the smallest distortion. The step size is always one. OTA stops when

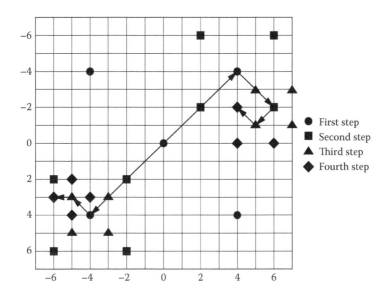

FIGURE 6.10
Cross-search algorithm.

the minimum distortion point is closeted between two points with higher distortion. The above procedure is repeated in the vertical direction about the point that has the smallest distortion in the horizontal direction. This search algorithm requires less time; however, the quality of the match is not very good.

6.16 Performance Comparison of Motion Search Algorithms

The performance of motion search algorithms is compared by both their prediction quality and computational complexity. With the same search window, only a full search (FS) achieves optimum prediction quality. All other fast-motion search algorithms have suboptimum prediction quality. However, they have a huge reduction in computational complexity compared to full search.

As a case study to compare the performance of a motion-search algorithm, analysis has been done using three standard video sequences in QCIF (176 × 144) format, each representing a different class of motion. These include Silent, News, and Foreman, presented in Figure 6.11. The first 100 frames of the above-mentioned sequences have been used for simulation.

In simulation the block distortion measure (BDM) is defined to be the mean square error (MSE). The block size is considered as 8 × 8 as trade-offs

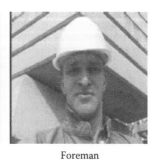

Silent   News   Foreman

FIGURE 6.11
Standard test video sequence.

between computational complexity and quality. The maximum motion in rows and columns is assumed to be ±7.

The quality of the reconstructed sequence should be estimated by subjective tests. One of the subjective metrics is mean square error (MSE), which is evaluated between the original frame and the reconstructed frame. The minimum value of MSE is better to predict quality.

Another widely used metric for comparing various image compression techniques is the peak signal-to-noise ratio (PSNR). The mathematical formula for PSNR is

$$PSNR = 10 \log_{10}\left(\frac{(2^b - 1)^2}{MSE}\right) \tag{6.7}$$

The b in Equation (6.7) is the number of bits in a pixel. For 8-bit uniformly quantized video sequence, $b = 8$. The higher the value of PSNR, the better is the quality of the compensated image.

Traditionally, computational complexity is measured by the number of search points being searched. For each search point, a BDM is being calculated. Theoretically, the computational complexity increases linearly with the number of points in a search window being searched. The number of search points being searched in FS is fixed. For a search window with the size ±7 pixels, the number of search points FS searches per block is $27 \times 2 + 1 = 225$. In practice, the blocks at the border areas of a frame require a fewer number of search points, because many of their search points are outside the border.

The average number of search points used per block for a fast-motion search can be compared with that used by FS. The reduction in the number of search points can be regarded as a theoretical speedup of that algorithm over FS.

Table 6.1 compares the performance of these algorithms. It is found that for all sequences FS shows higher PSNR values as compared to other algorithms with increased CPU time. Coarse searches in TDLS and 3SS can locate the rough position of the global minimum and subsequent four searches can find the best motion vector. They perform well in large motion sequence because search

TABLE 6.1

Performance of Block-Matching Algorithms

Algorithms	Average MSE	Silent Average PSNR	Average CPU Time (seconds)
FS	12.54	37.86	15.70
3SS	14.84	37.31	1.15
OTA	27.57	36.29	0.37
CSA	18.44	36.41	0.48
TDLS	15.97	37.11	1.43
		News	
FS	17.10	38.14	15.70
3SS	18.86	37.88	1.26
OTA	24.89	37.53	0.27
CSA	22.88	37.59	0.44
TDLS	20.05	37.91	1.39
		Foreman	
FS	26.36	34.22	15.70
3SS	34.13	33.27	1.11
OTA	42.49	31.59	0.50
CSA	50.98	30.79	0.52
TDLS	32.06	33.44	1.58

points are quite evenly distributed over the search window. For a higher motion Foreman sequence, the PSNR for TDLS is 33.44 and 33.27 for 3SS, which is higher than any other algorithm apart from FSA. However, TDLS required higher CPU time for computation as compared to other fast algorithms. OTA performs a one-dimensional gradient descending search on the error surface twice. Although it desires less computational time as compared with other fast block-matching algorithms, its prediction quality is low, which is reflected by the PSNR entries. This is because a one-dimensional gradient descend search is sufficient to provide a correct estimation of the global minimum position. In comparison with the other fast block-matching algorithms, while the computational complexity of the CSA is the second lowest, its compensation performance is not.

6.17 Video Coding Standards

Optimizing the coding efficiency is the main objective of the digital video coding standards.

Coding efficiency is the capability to retain video quality within the given specified bit rate.

The primary objective of the video coding standard is to offer freedom to the developer while designing and implementing the encoder and decoder module. This liberty in the development allows the standard to be adapted to a broad variety of platform architectures, application environments, and computing resource constraints. Liberty in the development is bounded by the condition of interoperability, which ensures that a video signal encoded by any vendor can be consistently decoded by others (Haskell and Puri 2012).

It is important to note that all video coding standards specify the syntax and semantics of the compressed bit stream produced by the video encoder and the decoding process to be followed at the decoder. However, standards do not mention the specific algorithms and procedure to be followed to create the required bit stream. The only necessity is that the encoder must follow the specified syntax to produce the resulting bit stream. Due to this, the quality of video codecs based on video standards mainly depends on the implementation of the encoder. Consequently, some implementations seem to produce better video quality than others.

Brief summaries of some of the video standards are discussed in the following sections.

Two standards bodies, the International Standards Organization (ISO) and the International Telecommunications Union (ITU), have developed a series of standards that have shaped the development of the visual communications industry. Both ITU-T and ISO have defined different standards for video coding. These standards are summarized in Table 6.2. Operating bit rates and the targeted applications differentiate the standards from each other. However, the standard supports a wide range of bit rates and hence they can be used for a variety of applications. A similar type of framework is adopted by the video coding standards for coding algorithms.

6.17.1 H.261 and H.263

To support video telephony and videoconferencing over ISDN circuit-switched networks, the ITU-T developed the H.261 standard (ITU-T 1990, 1993).

TABLE 6.2

Video Coding Standards

Organization	Standard	Typical Bit Rate	Typical Applications
ITU-T	H.261	$p \times 64$ Kbits/s, $p = 1, ..., 30$	ISDN Video phone
ISO	MPEG-1	1.2 Mbits/s	CD-ROM
ISO	MPEG-2	4–80 Mbits/s	SDTV, HDTV
ITU-T	H.263	64 Kbits/sec or below	PSTN Video phone
ISO	MPEG-4	24–1024 Kbits/s	Variety of applications

The standard was designed to offer computationally simple video coding for multiple 64 Kbit/s bit rates. A hybrid DPCM/DCT model with integer accuracy motion compensation is adopted by the standard. The ITU-T working group developed new standard H.263 (ITU-T 1990, 1993) with improved compression performance over H.261. This provides better compression than H.261, supporting basic video quality at bit rates of below 30 Kbit/s and is part of a suite of standards designed to operate over a wide range of circuit- and packet-switched networks.

6.17.2 MPEG-1 Video Coding Standard

The MPEG-1 standard is the first multimedia standard that provides specification regarding coding, compression, and transmission of audio and video data. The impetus of the standard was to store the data on CD-ROM with a high transfer rate. The quality of MPEG-1 is similar to VHS VCR (ISO/IEC JTC 1 1993).

6.17.3 MPEG-2 Video Coding Standard

The objective of the MPEG-2 standard was to offer the capability to compress, code, and transmit high-quality audio and video multimedia signals over broadband networks. The standard offers a testing feature to check the coded bit stream and decoder output. Testing is offered through software simulation at the encoder and decoder. MPEG-2 provides an error correction facility, which is suitable for transmission of a television signal. Video bit rates above 2 Mbit/s are supported by the standard (Haskell and Puri 2012). Further, MPEG-2 was extended to provide for a high-definition television signal.

6.17.4 MPEG-4

The initial goal of the MPEG-4 standard was to support low bit rate applications. Additional features were later added to support high-compression,

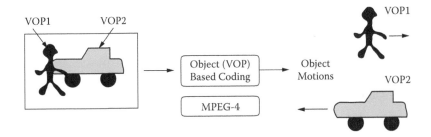

FIGURE 6.12
MPEG-4 VOP.

content-based interactivity and universal access (ISO/IEC JTC 1 1999, 2000, 2004; Srinivasan and Rao 1985). Figure 6.12 specifies the concept of content-based coding of MPEG-4. Every incoming image is decomposed into the number of randomly shaped regions known as video object planes (VOPs). Similar to the H.263 coding algorithm, each VOP is then coded. Arithmetic coding is applied to encode the shape of the VOP. MPEG-4 is different than the earlier standard in a way that it codes audio and video data. The earlier standard used to code video and corresponding audio data, MPEG-4 adopts to code every video and audio object in a frame. These separately coded audio and video objects are decoded and synchronized at the decoder. Several variants of MPEG-4 are presented based on different ways to represent and compress shape information (Haskell and Puri 2012).

6.17.5 MPEG-7

MPEG-7 is an ISO/IEC standard, previously named the *multimedia content description interface*. The standard was aimed to provide facility of retrieving multimedia information from the video signal. The standard supports some degree of interpretation of the multimedia information content in a video data. For example, a video may contain information in the form of text, images, and audio data. All these data are defined by a standard description by MPEG-7. As a result of this standard, fast and efficient retrieval of multimedia content is permitted. The applications of MPEG-7 can be categorized into pull-and-push scenarios. For the pull scenario, MPEG-7 technologies can be used for information retrieval from a database or from the Internet. For the push scenario, MPEG-7 can provide the filtering mechanism applied to multimedia content broadcast from an information provider.

6.17.6 H.264/MPEG-4 Part 10/AVC

The aim of H.264/MPEG-4 part 10/AVC standard was to increase the bit-rate of earlier video coding standards without losing the quality (Liu and Zaccarin 1995). The standard was originally known as H.26L or JVT, for the joint video team, in which the ISO and ITU organizations worked together to complete the standardization. H.264 is the ITU name for the standard; MPEG-4 part 10, advanced video coding (AVC) is the ISO name. Due to the high compression quality, H.264 is accepted by many video coding applications ranging from the iPod to TV broadcasting standards.

6.17.7 Scalable Video Coding

The joint video team (JVT) of ITU-T Video Coding Experts Group (VCEG) and ISO/IEC Moving Picture Experts Group (MPEG) jointly developed the scalable extension of H.264/AVC (SVC). The number of applications of video processing is driven by the corresponding progress in the related technology (e.g., coding, network, memory, and processing power) (ISO/IEC JTC 1 1999, 2000, 2004; Liu and Zaccarin 1995). It is possible to transmit video signals not only over wired channels but also on wireless channels with variable bandwidth. It is promising that we can store the video data on low-end memories as well as on high-capacity memories. The display devices used are as small as screens of mobile phones to the high-resolution screens of monitors. Conventional video systems were guaranteed to have required bandwidth. It was then necessary to compress the video data as per the available bandwidth. In today's era of Internet transmission, the encoder does not have information about available bandwidth and it is not fixed. In a conventional video coding system, the bit rate is being fixed; hence, it is not applicable for such kinds of applications. Scalable video coding is the solution for such applications which supports variable bit-rate capacity.

There are different layers in the process of scaling. Some minimum number of bits is necessary to maintain a certain minimum quality; this bit stream is called the base layer. Further, there can be additional bit streams that can be added to enhance this basic quality. This is termed *enhancement layers*. With these enhancement layers it is possible to construct high-resolution video signals. This gives freedom to achieve required quality.

The earlier standards MPEG-2 Video/H.262 (ISO/IEC JTC 1 1994) and MPEG-4 Visual (ISO/IEC JTC 1 2003, 2004, 2005) do have a scalable tool; however, these features could not have been utilized due to the characteristics of traditional video transmission systems where scalabilities are not really required. And second, a scalable system loses encoder efficiency and the complex architecture of the decoder must be maintained.

MPEG and ITU-T Video Coding Experts Group (VCEG) (ISO/IEC JTC 1 2007) together published a scalable extension of H.264/MPEG-4 AVC (Advanced Video Coding) which makes the scalable extension the state-of-the-art scalable video codec.

6.17.8 Next-Generation Video Compression—HEVC

MPEG and ITU have recently approved a new video-compression standard known as high-efficiency video coding (HEVC), or H.265, that is set to provide double the capacity of today's leading standards (Sullivan, Ohm, and Han 2012).

One of the primary target areas for HEVC compression is high-resolution video, such as HD and UHD.

HEVC has the same basic structure as previous standards such as MPEG-2 video and H.264. However, HEVC contains many incremental improvements such as

- More flexible partitioning from large to small partition sizes
- Greater flexibility in prediction modes and transform clock sizes

- More sophisticated interpolation and deblocking filters
- More sophisticated prediction and signaling of modes and motion vectors
- Features to support efficient parallel processing

The result is a video coding standard that can enable better compression at the cost of potentially increased processing power. With HEVC it should be possible to store and to transmit video more efficiently than with earlier technologies such as H.264 (Ohm, Sullivan, and Schwarz 2012).

References

N. Ahmed, T. Natrajan, and K. R. Rao, Discrete cosine transform. *IEEE Transactions on Computers*, C-23(1): 90–93, December 1984.

R. J. Clarke, *Transform Coding of Images (Microelectronics and Signal Processing)*, Academic Press, London, 1985.

M. Ghanbari, The cross-search algorithm for motion estimation. *IEEE Transactions on Communication*, 38(7): 950–953, July 1990.

P. Gorpuni, Development of Fast Motion Estimation Algorithms for Video Compression. Thesis submitted to Master of Technology in Telematics and Signal Processing Department of Electronics and Communication, National Institute of Technology Rourkela, 2009, pp. 11–12.

B. G. Haskell and A. Puri, MPEG Video Compression Basics, L. Chiariglione (ed.), *The MPEG Representation of Digital Media*, DOI 10.1007/978-1-4419-6184-6_2, SpringerScience+Business Media, New York, 2012.

ISO/IEC JTC 1, Advanced video coding for generic audio-visual services, *ITU-T Recommendation H.264 and ISO/IEC 14496-10 (MPEG-4 AVC)*, Version 1: May 2003, Version 2: January 2004, Version 3: September 2004, Version 4: July 2005.

ISO/IEC JTC 1, Advanced video coding for generic audio-visual services, *ITU-T Recommendation H.264* Amendment 3, ISO/IEC 14496-10/2005: Amd 3, Scalable extension of H.264 (SVC), July 2007.

ISO/IEC JTC 1, Coding of audio-visual objects—Part 2: Visual, *ISO/IEC 14496-2 (MPEG-4 Visual)*, Version 1: April 1999, Version 2: February 2000, Version 3: May 2004.

ISO/IEC JTC 1, Coding of moving pictures and associated audio for digital storage media at up to about 1.5 Mbit/s—Part 2: Video, *ISO/IEC 11172-2 (MPEG-1 Video)*, March 1993.

ISO/IEC JTC 1, Generic coding of moving pictures and associated audio information—Part 2: Video, *ITU-T Recommendation H.262 and ISO/IEC 13818-2 (MPEG-2 Video)*, November 1994.

ITU-T, Video codec for audiovisual services at $p \times 64$ Kbit/s, ITU-T Recommendation H.261, Version 1: November 1990, Version 2: March 1993.

ITU-T, Video coding for low bit rate communication, *ITU-T Recommendation H.263*, Version 1: November 1995, Version 2: January 1998, Version 3: November 2000.

J. R. Jain and A. K. Jain, Displacement measurement and its application in interframe image coding. *IEEE Transactions on Communications*, 29(12): 1799–1808, December 1981.

T. Koga and Ishiguro, Motion compensated inter-frame coding for video conferencing. *Proceedings of National Telecommunication Conference*, New Orleans, pp. G5.3.1–G5.3.5, December 1981.

B. Liu and A. Zaccarin, New fast algorithms for the estimation of block motion vectors. *IEEE Transactions on Circuits and Systems Video Technology*, 3: 440–445, December 1995.

J. -R. Ohm, G. J. Sullivan, and H. Schwarz, Comparison of the coding efficiency of video coding standards—Including high efficiency video coding (HEVC). *IEEE Transactions on Circuits and Systems for Video Technology*, 22(12): 1669–1684, December 2012.

I. E. G. Richardson, *H.264 and MPEG-4 Video Compression, Video Coding for Next-Generation Multimedia*, Robert Gordon University, Aberdeen, UK, 2003.

T. Sikora, *Digital Video Coding Standards and Their Role in Video Communications, Signal Processing for Multimedia*, J. S. Byrnes (Ed.), IOS Press, 1999, pp. 225–252.

T. Sikora, MPEG digital video coding standards. *IEEE Signal Processing Magazine*, pp. 82–100, September 1997.

R. Srinivasan and K. R. Rao, Predictive coding based on efficient motion estimation. *IEEE Transactions on Communications*, 33(8): 888–896, August 1985.

G. J. Sullivan, J. -R. Ohm, and W. -J. Han, Overview of the high efficiency video coding (HEVC) standard. *IEEE Transactions on Circuits and Systems for Video Technology*, 22(12): 1649–1667, December 2012.

Y. Wang, J. Ostermann, and Y. Q. Zhang, *Video Processing and Communications*, Tsinghua University Press and Prentice Hall, Beijing, China, 2002.

M. Wien, H. Schwarz, and T. Oelbaum, Performance analysis of SVC. *IEEE Transactions on Circuits and Systems for Video Technology*, 17(9): 1194–1203. September 2007.

W. Yao, Z. Guo, and L. Susanto Rahardja, The scalable video coding in a nutshell. *Synthesis Journal*, section 4, pp. 89–108, 2008.

7

Image Quality Assessment

With the rapid growth in Internet technologies and the enormous use of multimedia content for different applications, from education to entertainment, paramount importance is given to the use of images and videos. In order to save space and transfer in less time, these images and videos are compressed. Due to compression, artifacts are generated, which degrades the quality of the images. Research is still going on to increase the compression ratio by degrading quality to a small extent. The extent to which quality is degraded should be analyzed for further performance improvement of the compression algorithm. The full-reference quality metrics such as peak signal-to-noise ratio, mean square error, structural similarity, and image fidelity are explained. If the reference image is not available, the technique used to determine quality is discussed. The image quality assessment technique based on information from the reference image (reduced reference) is also discussed.

7.1 Introduction

During the last three decades, because of advances in imaging technologies and the rapid growth of the Internet, digital images and videos are now widely used for representing information in entertainment. We have studied in previous chapters that due to the large size of images and videos, they need to be compressed. As a result of the compression artifacts, the quality of images degrades. The quality also suffers during acquisition, transmission, and reproduction and processing. So, image quality measurement has a wide importance in many applications related to image processing. Efficient and automatic objective quality evaluation using the development of quality assessment algorithms is the main goal of quality assessment research. Also, it should be consistent with subjective/human assessment. Identification and quantification of image quality degradations by the algorithms for the assessment are also necessary. This will be to update the performance of the compression process or channel.

Human beings are the final recipients in many image-processing applications; images are assessed by subjects (observers). This method is based on the mean of the opinion given by observers. Even though correct judgment is possible, this method is fairly expensive and slow. This approach is called *subjective image quality assessment* (IQA).

Predication of the perceived image quality accurately based on computational models is another way. The prediction is supposed to be in correlation with subjective assessment. This approach is called an objective IQA.

Thus, image quality assessment techniques are classified mainly into (1) subjective and (2) objective. Further objective quality assessment is classified into the following:

1. *Full reference (FR)*—In this type of assessment, the original image before processing (compression) is used as reference. The decompressed or distorted image is compared with reference, typically using distance criteria. This method is still used by the researchers to evaluate performance of the compression scheme developed.

2. *No reference (NR)*—In this approach, there is an image available as reference. Certain artifacts/features in an image like blockiness and ringing are measured and given to a well-trained prediction model to get the quality score. The model uses subjective assessment during training. The task is difficult but is drawing the attention of many researchers to develop no reference quality metrics.

3. *Reduced reference (RR)*—Certain features like edge information and pixel values at a fixed position are extracted and attached as side information along with the compressed image. For assessment, the features are compared and reduced so that a reference quality metric can be developed. This approach can also be used for image quality improvement (repair).

Subjective IQA is discussed in the next sections, and all objective IQA techniques are discussed in further sections along with MATLAB programs and comparison between various quality metrics.

7.2 Subjective Image Quality Analysis

Subjective image quality analysis may be the only best method for quantifying visual image quality. The mean opinion score (MOS) is a widely used method for subjective assessment of image quality. In this method, five or more grade scales are used. Experts and nonexperts are asked to give grades for images as 5 for "Excellent," 4 for "very good," 3 for "good," 2 for "bad," and 1 for "very bad." The average of the grades given to that image is calculated. In many cases, grades out of 10 are used instead of 5. MOS is also calculated using the equation shown below:

$$MOS = \sum_{i=1}^{5} i * P(i) \qquad (7.1)$$

where i is grade, and $P(i)$ is the probability of the grade.

Evaluation by subjective methods is usually very inconvenient, expensive, time-consuming, and not suitable for real-time applications. Using MOS, no reference quality metrics are developed.

7.3 Objective Image Quality Assessment

Mainly, objective quality assessment is used as follows:

1. They are used in control systems to monitor image quality. In such systems images captured or processed are assessed, and if they do not have some measured value they are recaptured or reprocessed with new settings of parameters. In video transmission systems, based on the quality of the compressed video, the allocation of bit streams is done, as shown in Figure 7.1.

2. They are used as benchmarks in image processing systems and algorithms, like compression. They are widely used in research in order to compare different algorithms.

3. The system parameter settings used in image-processing and transmission systems can be optimized. In such systems, the metrics are used for judgment of quality as well as minimizing error (image repair).

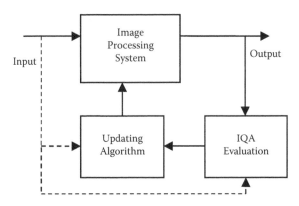

FIGURE 7.1
Image quality assessment system.

7.4 Full-Reference Image Quality Assessment

If an original image is available for the assessment, then the comparison between the distorted image and the original image is carried out, the method is called full reference (FR). This method is widely used to evaluate the performance of the algorithms and techniques used in compression. However, the metrics or measures developed may not give accurate results.

Eskicioglu and Fisher (1995) have used various image quality metrics based on full reference. The classification of the image quality measures in different groups based on properties is done by them. The classification is shown in Figure 7.2. These are normally used to assess the performance of compression algorithms.

7.4.1 Peak Signal-to-Noise Ratio, Mean Square Error, and Maximum Difference

The simplest, traditional metric for measuring image quality is the mean square error (MSE). Suppose $x(m,n)$ and $\hat{x}(m,n)$ are original (perfect quality) image and reconstructed (distorted) image, respectively, and the size of the image is $M*N$. Then, the value of MSE can be calculated as

$$MSE = \frac{1}{MN}\sum_{m=1}^{M}\sum_{n=1}^{N}(x(m,n)-\hat{x}(m,n))^2 \qquad (7.2)$$

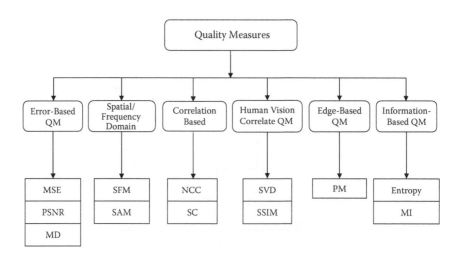

FIGURE 7.2
Group of quality measures for image compression.

Here, the difference of pixel values at coordinates is taken. If the original image and the reconstructed image are very close, then its value is almost zero.

The ratio of the maximum possible power of a signal and the power of noise that affects the original form of the signal is called the *peak signal-to-noise ratio* (PSNR). The PSNR value approaches infinity as the MSE approaches zero, meaning that an excellent image quality has a high value of PSNR. PSNR is calculated as follows:

$$PSNR = 10\log\frac{L^2}{MSE} \tag{7.3}$$

where L is the dynamic range, and for an eight-bit image the range of pixel values is 0 to 255. In this case, L is taken as 255.

Typical values of the PSNR range from 25 to 45 dB. A higher value indicates very good quality.

The MSE is defined as

$$E_p = \sum_{m=1}^{M}\sum_{n=1}^{N}(x(m,n) - x^{m.n})^p \tag{7.4}$$

where $p\varepsilon\ [1, \infty)$.

If $p = 1$, this results in mean absolute error, whereas if $p = 2$, its MSE $p = \infty$ will give the maximum absolute difference, also called *maximum difference* (MD), and presented as

$$MD = Max(\,|x(m,n) - \hat{x}(m,n)|\,) \tag{7.5}$$

For all compression techniques, the MD is highly correlated with MOS. It is most suitable for full-reference IQA, being a very simple measure. The larger the value of MD, the poorer is the image quality. All these measures are simple and easy to compute. Since MSE refers to energy of the signal and is preserved after applying unitary transformations, it is widely used.

7.4.2 Why Mean Square Error Is Not Always Correct

The image of Barbara is corrupted by noise and blur, as shown in Figure 7.3. Image Figure 7.3b is of higher visual quality than Figure 7.3d, but the MSE value is still the same for all (i.e., 252.25).

The PSNR will be the same even though the picture quality is different. If the first two rows and columns are removed and the image is shifted with replicating the last two rows and columns, then the image looks nearly the same as the original, but the PSNR is dropped to a large extent. The MSE value has no effect of change in sign of the error signal pixel as the perception

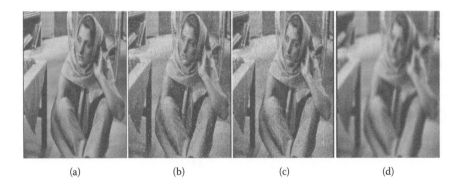

(a)   (b)   (c)   (d)

FIGURE 7.3
Barbara images. (a) The original image. (b) The image with random noise. (c) The image with Gaussian noise. (d) The image after averaging blur.

of the quality of the image is estimated by the absolute magnitude of the error signal.

As a result, for the same error or noise, the distortion measure remains the same. As discussed, if perceived image quality is good and PSNR is low, this quality metric fails. In such a case, the image quality is judged based on the HVS parameters such as contrast, luminance, and structural information. Wang and Bovik (2004) designed a metric called the mean structural similarity index measure (MSSIM), which performs well in such situations and correlates well with perceived image quality by HVS.

7.4.3 Spectral Activity Measure and Spatial Frequency Measure

The spatial frequency measure (SFM) indicates the degree of overall activity in an image, defined by R (row frequency) and C (column frequency):

$$SFM = \sqrt{\left(R^2 + C^2\right)} \tag{7.6}$$

$$R = \sqrt{\left[\frac{1}{MN} \sum_{m=1}^{M} \sum_{n=2}^{N} [x(m,n) - x(m,n-1)]^2\right]} \tag{7.7}$$

$$C = \sqrt{\left[\frac{1}{MN} \sum_{m=1}^{M} \sum_{n=2}^{N} [x(m,n) - x(m-1,n)]^2\right]} \tag{7.8}$$

As per the equations, the differences are between next and present pixels, row and column wise. The SFM value for the test image baboon (mandrill) is large because the image has a lot of detail—that is, large variations in pixel values. A larger SFM means the image contains higher frequencies.

In the spectral domain, the spectral activity measure (SAM) is defined as a metric of image predictability. The evaluation deals with the DFT coefficients of an image. The dynamic range of SAM is $[1,\infty]$. The larger the values of SAM, the higher will be the predictability:

$$SAM = \frac{\frac{1}{MN}\sum_{m=1}^{M-1}\sum_{n=0}^{N-1}|X(m,n)|^2}{\left[\prod_{m=0}^{M-1}\prod_{n=0}^{N-1}X(m,n)^2\right]^{1/MN}} \tag{7.9}$$

where $X(m,n)$ is the discrete Fourier transform of the image. The numerator is the arithmetic mean of the square of the Fourier coefficients in the image and the denominator is the geometric mean of the Fourier coefficients. The smaller the value of SAM, the lower will be the predictability.

While comparing the original image and distorted images, their SFM and SAM values are compared, respectively. These measures are widely used in analyzing the performance of compression algorithms, especially for high spectral activity images.

7.4.4 Normalized Cross-Correlation Measure

In image processing, the range of intensity values of pixels is changed by normalization, also referred to as dynamic range compression.

The degree of similarity between two images is often calculated with the correlation between two images (cross-correlation), mathematically defined as given below:

$$NCC = \frac{\sum_{m=1}^{M}\sum_{n=1}^{N}\left(x(m,n)*\hat{x}(m,n)\right)}{\sum_{m=1}^{M}\sum_{n=1}^{N}[x(m,n)]^2} \tag{7.10}$$

$x(m,n)$ and $\hat{x}(m,n)$ are original image and reconstructed image, respectively. This measure is complementary to the difference measure. When two images are similar, then the value of NCC approaches 1.

7.4.5 Structural Content and Image Fidelity

This measure compares the structure of the image. The weights of the pixels are compared position wise. If they are the same, the ratio becomes 1. During distortion, the pixel values at a particular position change, for example, edge distortion, and then the ratio drops. If two images are similar, then the SC value is 1. If the value is very much less than 1, then the image quality is poor. The SC is calculated as

$$SC = \frac{\sum_{m=1}^{M}\sum_{n=1}^{N}x(m,n)^2}{\sum_{m=1}^{M}\sum_{n=1}^{N}\hat{x}(m,n)^2} \tag{7.11}$$

Natural images are supposed to be very structured.

Another better stable metric is image fidelity and is given by

$$IF = 1 - \sum_{j=1}^{M}\sum_{k=1}^{N}[x(m,n)-\hat{x}(m,n)^2] \Big/ \sum_{j=1}^{M}\sum_{k=1}^{N}[x(m, n)^2] \tag{7.12}$$

The value is in between 0 and 1. If closer to 1, that means that the reference and degraded images are similar.

7.4.6 Mean Structural Similarity Index

To extract structural information from images, HVS is highly adapted. Measurement of structural similarity (or distortion) is well correlated with perceptual image quality. Figure 7.4b,c shows images with enhancement in contrast and brightness, and the structural information in such images is not changed, whereas in the JPEG compressed image shown in Figure 7.4d, the structure has been changed.

The structural similarity index metric (SSIM) is the measure of structural changes in image (luminance, contrast, and other errors). For original signal x, and the reconstructed signal y, the SSIM index is defined as

$$SSIM(x,y) = [l(x,y)]^\alpha [c(x,y)]^\beta [s(x,y)]^\gamma \tag{7.13}$$

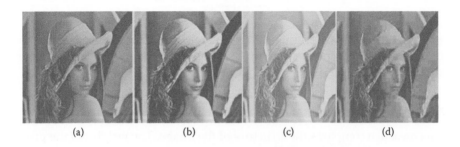

| | | | |
| (a) | (b) | (c) | (d) |

FIGURE 7.4
(a) Original image of Lena. (b) Contrast enhanced. (c) Brightness enhanced. (d) JPEG compressed with very low quality.

It is assumed that in SSIM the value of α, β, *and* γ is equal to one. The luminance, contrast, and structural components are shown below:

$$l(x,y) = \frac{2\mu_x\mu_y + C1}{\mu_x^2 + \mu_y^2 + C1}$$

$$C(x,y) = \frac{2\sigma_x\sigma_y + C2}{\sigma_x^2 + \sigma_y^2 + C2} \qquad S(x,y) = \frac{\sigma_{xy} + C3}{\sigma_x\sigma_y + C3} \tag{7.14}$$

where μ_x, μ_y are the statistical means of x and y, respectively. And σ_x, σ_y are standard deviations and σ_{xy} is given by

$$\sigma_{xy} = \frac{1}{N-1}\sum_{i=1}^{N}(x - \mu_x)(y - \mu_y) \tag{7.15}$$

where N is the total number of pixels. The constant C1 avoids instability caused when $\mu_x^2 + \mu_y^2$ is very close to zero. The mean of SSIM is MSSIM.

The value of MSSIM is in the range of 0 to 1. The MSSIM performs better than PSNR. The MSSIM can also be designed in the frequency domain using continuous wavelet transform. Readers are referred to Wang and Bovik (2006) for further reading. If C1 and C2 are equated to zero, then the metric is called the universal quality index (UQI).

7.4.7 Singular Value Decomposition Measure

Another important measure used in subjective evaluation is based on the type and amount of distortion and the distribution of the error in decompressed images. The human visual system should ideally be imitated by an image quality measure. Singular value decomposition (SVD) is a widely used image quality measure for grayscale images. A wide range of distortions can be predicted by using a graphical or scalar measure such as SVD. The SVD of X is calculated as follows.

Let $X = USV^T$, where X is the real matrix and U and V are orthogonal matrices. And $U^T U = I$, $V^T V = I$, and $S = diag(s1, s2,...)$ are the singular values of X.

The columns of U show the left singular vectors of X and the columns of V show a right singular vector of X. SVD is a widely used tool in many signal processing applications. The discussion here is applicable to gray images only. For the color images, the operations are performed on luminance channel Y. There are two methods of SVD calculation: global (on full image) and local (8 × 8 blocks). Local errors are obtained on smaller blocks. They are averaged to obtain the global measure. This is presented by

$$D_i = sqrt\left[\sum_{i=1}^{n}\left(s_i - \hat{s}_i\right)^2\right] \tag{7.16}$$

where s_i are singular values of the original block, $\hat{s}_i$ are singular values of the distorted block, and n is the block size. For image size P, we have $(P/n) \times (P/n)$ blocks. A distortion map represents the set of distances in a graph. The block size used is 8 × 8.

7.4.8 Edge-Based Pratt Measure

Pratt (1978) introduced this measure, which considers both the accuracy of the missing edge elements and edge location. Pratt had defined a figure of merit as

$$PM = \frac{1}{\max\{n_d, n_t\}}\sum_{i=1}^{n_d}\frac{1}{1 + ad_i^2} \tag{7.17}$$

where n_d is the number of detected edge points, n_t is the number of ground truth edge points, and d_i is the distance of the closest edge for the detected ith edge pixel. The empirical calibration constant is a, and its recommended value by Pratt is 1/9. *Ground truth* is nothing but the binary edge field obtained from the uncompressed image. The factor $\max\{n_d, n_t\}$ punishes the number of edges that are missing. The relative comparison between noisy edges and thin but offset edges is provided by this scaling factor.

PM indicates edge quality. It is the representation of the distances between the edges. It is a relative measure; hence, its range varies in [0, 1], where 1 represents the optimal value (i.e., the edges detected coincide with the ground truth).

In recent times, Chen et al. (2006) presented an edge-based measure called ESSIM that is edge-based SSIM. They replaced structural information in the SSIM by edge information calculated by applying the Sobel edge operator on an original and a reconstructed image and finding the amplitude and direction of the edges and comparing them.

7.4.9 Entropy

One of the widely used measures for the information content of images is the entropy. Initially, Shannon proposed it for measuring the information content per symbol, arriving from a random information source.

The measure of source information is identical to the measure of image information. To measure this information from the image, Shannon's theorem must be satisfied. So in image analysis, Shannon's entropy can be used and is given by a formula

$$H(X) = -\sum_{i=1}^{I} P(x_i) log P(x_i)$$

(7.18)

where i is the number of gray level of the image's histogram with range [0–255] and $P(x_i)$ is the probability of occurrence. For assessing quality, the entropies of the original and the reconstructed images are calculated.

7.4.10 Mutual Information

Mutual information (MI) is defined as a measure of uncertainty and is given by

$$MI\ (X, Y) = H\ (X) + H\ (Y) - H\ (X, Y)$$

(7.19)

When the mutual information of two images is maximum, it is the optimum transformation.

7.5 No Reference Image Quality Assessment

As discussed in Section 7.1, because processed images are sent over the network or stored independently, the quality assessment without any reference is a major challenge. All the quality measures discussed except no reference (NR) use mathematical computations and also have lower computational complexity, but have a serious drawback that they do not always give correct assessment. Being statistically based, these metrics may indicate proper values but the images may not be good as far as human vision is concerned. This has been proved through various experiments based on image quality assessment. The basic theme in NR image quality is the exploration of information fidelity or quantification of loss of information due to the distortion process (compression, noise, etc.). All these metrics were parametric based. Some are also based on image type, like JPEG.

The quality metric that calculates the image quality in the absence of reference or original image that closely correlates with human perception is supposed to be a Herculean task. This is mainly due to the restricted know-how of the HVS.

Feature Vector

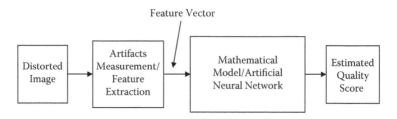

FIGURE 7.5
Block diagram presentation of the generalized method of NR quality score determination.

With prior knowledge about the degradation, it is possible to do effective NR quality assessment. The basic philosophy behind NR IQA which presents assessment in terms of quality score in the range of 0 to 10 similar to that of MOS is as shown in Figure 7.5.

An important aspect here is measurement of distortion or artifacts generated during compression or extraction of certain parameters in the image so as to form a feature vector. Using these measurements or feature vector, a mathematical model, probability-based model, or neural network model is trained using MOS as a target. Once properly trained, it is tested for any input image and a quality score is obtained, which is correlated with HVS/MOS.

Techniques available in NR metric designs are varied and can be divided into three broad categories:

- *Distortion-specific approach*—This algorithm measures a specific distortion type on an individual basis and provides the quality score accordingly. The NR image quality assessment algorithm examples are based on blockiness, blur estimation, and ringing artifact measurement.

- *Feature extraction and learning approach*—In this, features from images are extracted and a neural network or any other classifier is trained to find a difference between distorted and undistorted images based on extracted features. A similar approach also uses a support vector machine (SVM).

- *Natural scene statistics (NSS) approach*—This assumes that a subspace of the entire space of possible images is occupied by natural or distorted images and tries to calculate the difference between the subspace of the natural image and the distorted image. It helps in the study of change in statistics of the distorted image. The drawback of the first approach is that it is application specific. An application-specific NR image quality assessment system is one that is specifically designed to handle a specific artifacts type and that is unlikely to be able to handle the other type of distortion. The second technique is superior like the feature extraction. Reliability of quality assessment is directly proportional to the number of features extracted. Though a natural scene statistic is a better approach, it requires extensive statistical modeling, and accurate robust generalization of the model is required. From the relevant literature, it is noticed that the NR quality measures are designed either by parametric or nonparametric methods or by using one or two distortion artifacts. They are also for specific types of images such as JPEG. The classifiers used are also limited.

Interested readers may refer to the References or Bibliography related to NQ IQA cited in this chapter for the details of parameters and features used.

7.6 Reduced-Reference Image Quality Assessment

In reduced-reference (RR) image quality assessment, features like edge, structure, and so forth, represent partial reference images. Accuracy in estimating quality and the amount of information required to describe the reference is compromised in RR as compared to FR and NR metrics. The selection of RR features depends upon the trade-off between data rate or size of the RR features and quality score prediction accuracy. At the receiver side, the FR method can be applied if a sufficient data rate is available to deliver additional information as part of the reference signal. The least data rate can represent the maximum image feature information. In addition to the above, extracted RR features provide information about the correction factor to be added to the distorted image to improve the quality of the image and in turn the quality score. The side information used for RR should be within 5% to 10% of the limits of the file size.

Figure 7.6 shows the process of RR quality assessment. Here, the extracted features are either sent to the receiver over an ancillary channel or can be attached as part of the compressed image. Features extracted at the receiver are compared with the features of the reference image. The quality score is computed based on the difference between the same. Many times using the difference, the features can be corrected in order to repair the loss or distortion. The features are extracted in spatial as well as in frequency domain.

In existing RR IQ algorithms, there are three different but related types of approaches:

- *Image distortions modeling-based approaches*—Application-specific environments use this approach. For instance, a set of typical distortion artifacts like blurring, blocking, and ringing can be identified when a standard image/video compression scheme is known. Then image features suitable to quantify these artifacts can be defined.

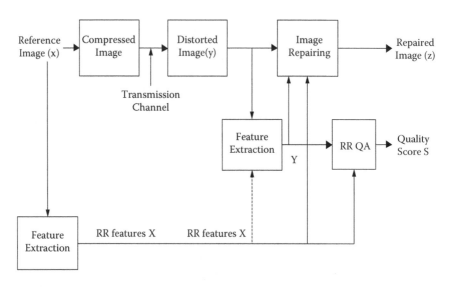

FIGURE 7.6
Process of RR-IQA.

- *HVS modeling-based approaches*—This approach provides a reduced description of the image. For this, perceptual features of low-level vision and its computational models are derived. Perceptual features are not directly related to any specific distortion type. Hence, this approach can be extended for general-purpose applications. A variety of distortion-specific RRIQA algorithms can be built under the same general framework.

- *Natural image statistics modeling-based approaches*—In this approach, image quality degradation can be quantified by modeling statistics of natural image. The fitting model is used as RR features. As this approach does not require any training with a low RR data rate, it is successful in performance testing for various image distortion types.

7.7 Concluding Remarks

The main focus of this chapter is full-reference quality measures. To compute the performance of a compression algorithm, developed metrics are widely used. The drawback of MSE is that it requires researchers to search and use other relatively stable quality metrics such as IF, SC, and MSSIM. The present research area is no reference quality assessment; the philosophy behind NR quality score determination is described. The use of reduced reference quality metrics will be mainly used for error correction in order to improve the quality of the image. The side information used for RR should be within 5% to 10% limits of the file size. Case studies for evaluation of a vector quantization–based image compression scheme and SPIHT-based image compression are described.

7.8 MATLAB Programs

1. Following is a MATLAB program for calculating MSE and PSNR. Readers can use their images with proper paths and compute the results.

```
x = imread ('xxxx.xxx') % read original or reference image
y = imread ('xxxx.xxx')% read distorted image
Diff = double(x)-double(y);        % difference of two images
diffsq = diff.*diff;               % Square of difference
diffsum = sum(sum(diffsq));        % Summation of difference
MSE = diffsum/(M*N); % Here size of image taken to be MxN, readers can
                 substitute their %image size
r = 65025/MSE;                     % square of 255
PSNR = 10*log10(r);
sprintf('The MSE is%f',MSE)        % Display MSE
sprintf('The PSNR is%f dB',PSNR)   % Diplay PSNR
```

There is a direct command in the MATLAB wavelet toolbox to compute MSE, PSNR, and maximum error. The syntax is [psnr,mse,maxerr] = measerr(x, y);%, *x* and *y* are reference and distorted image, respectively.

2. The program for finding MSSIM and UQI is as follows:

```
x = imread ('xxxx.xxx') % read original or reference image
y = imread ('xxxx.xxx')% read distorted image
K(1) = 0.01;                       % default settings
K(2) = 0.03;                       %
L = 255;
C1 = (K(1)*L)^2;
C2 = (K(2)*L)^2;
im1 = double(x);
im2 = double(y);
mu1 = mean2(im1); % Mean Calculations
mu2 = mean2(im2);
mu1_sq = mu1.*mu1;
mu2_sq = mu2.*mu2;
mu1_mu2 = mu1.*mu2;
sigma1 = std2(im1);                % Standard deviation calculations
sigma2 = std2(im2);0
```

```
sigma1_sq = sigma1.*sigma1;
sigma2_sq = sigma2.*sigma2;
diffx1 = im1(:,:,1)-(ones(M,N)*mu1); % Substitute size of
                                       image M rows and N
                                       columns
diffx2 = im1(:,:,2)-(ones(M,N)*mu1);
diffx3 = im1(:,:,3)-(ones(M,N)*mu1);
diffx = (diffx1+diffx2+diffx3)/3;
diffy1 = im2(:,:,1)-(ones(M,N)*mu2);
diffy2 = im2(:,:,2)-(ones(M,N)*mu2);
diffy3 = im2(:,:,3)-(ones(M,N)*mu2);
diffy = (diffy1+diffy2+diffy3)/3;
sigma12 = diffx.*diffy;
ssim_map = ((2*mu1_mu2 + C1).*(2*sigma12 + C2))./((mu1_sq +
mu2_sq + C1).*(sigma1_sq + sigma2_sq + C2));
mssim = mean2(ssim_map);
%%%%%%%%%%%%%%%%%Universal Quality Index Calculations%%%%%%%%%
C1 = 0;C2 = 0;
UQI = ((2*mu1_mu2 + C1).*(2*sigma12 + C2))./((mu1_sq + mu2_sq
+ C1).*(sigma1_sq + sigma2_sq + C2));
UQI = mean2(UQI);
```

3. The MATLAB program for computing average difference (AD), maximum difference (MD), structural content (SC), and image fidelity (IF) is as follows:

```
x = imread ('xxxx.xxx') % read original or reference image
y = imread ('xxxx.xxx')% read distorted image
x = double(x);
y = double(y);
% calculation of Average Difference (AD)
adiff = x-y;
ad = sum(adiff);
ad = sum(ad);
ad = ad/(M*N);
ad = (ad(:,:,1)+ad(:,:,2)+ad(:,:,3))/3;
sprintf('The average Difference(AD)is =%f',ad)
% Calculation of Maximum Difference
MDD = max(max(adiff));
MD = (double(MDD(:,:,1))+double(MDD(:,:,2))+
double(MDD(:,:,3)))/3;
MD = uint8(MD);
sprintf('The maximum difference is%d.',MD)
% Calculation of Structural Content (SC)
num = x.*x;
num = sum(sum(num));
den = y.*y;
den = sum(sum(den));
SC1 = num(:,:,1)/den(:,:,1);
```

```
SC2 = num(:,:,2)/den(:,:,2);
SC3 = num(:,:,3)/den(:,:,3);
SC = (SC1+SC2+SC3)/3;
sprintf('The structural content is%f',SC)
% Calculation of normalized cross correlation
N = x.*y;
N = sum(sum(N));
D = x.*x;
D = sum(sum(D));
NK1 = N(:,:,1)/D(:,:,1);
NK2 = N(:,:,2)/D(:,:,2);
NK3 = N(:,:,3)/D(:,:,3);
NK = (NK1+NK2+NK3)/3;
sprintf('The normalized cross correlation is%f',NK)
% Calculation of Image fidelity
N2 = sum(sum((x-y).*(x-y)));
D2 = sum(sum(x.*x));
R1 = N2(:,:,1)/D2(:,:,1);R2 = N2(:,:,2)/D2(:,:,2);R3 =
N2(:,:,3)/D2(:,:,3);
RT = (R1+R2+R3)/3;
IF = 1-RT;
sprintf('The image fidelity is%f',IF)
```

Note that for calculating entropy, there is a direct command in MATLAB. Entropy (I) will calculate the entropy of image I.

7.9 Case Studies

Dandawate and Joshi (2009) developed a vector quantizer called the generic VQ. The performance analysis of the VQ compression is evaluated using quality metrics. The following chart shows the performance analysis.

Images	PSNR	SC	IF	MSSIM
Peppers	28.16	1.00	0.91	0.98
Cameraman	25.01	1.01	0.80	0.97
Fingerprint	23.22	1.01	0.70	0.97
CT scan	23.52	1.03	0.70	0.98
Lena	27.87	1.00	0.91	0.98
Tiffany	25.85	1.01	0.82	0.95
Bird	33.62	1.00	0.97	0.99
Peppers	34.49	1.00	0.97	0.99
Lighthouse	22.99	1.02	0.77	0.95
Mandrill	22.56	1.02	0.74	0.89
Bridge	23.69	1.01	0.78	0.95

Images	PSNR	SC	IF	MSSIM
X-ray	27.68	1.00	0.84	0.99
Rings	23.23	1.03	0.88	0.99
MRI	23.99	1.02	0.78	0.98
Crowd	24.34	1.02	0.81	0.95
City	21.62	1.01	0.76	0.85
Rose	26.74	1.02	0.73	0.98
Letters	18.96	1.03	0.85	0.86

The images are from data from the Computer Vision Web page (www.image-processing.com). The size chosen for the experimentation is 256 × 256.

References

G. -H. Chen, C. -L. Yang, L. -M. Po, and S. -L. Xie, Edge-based structural similarity for image quality assessment. In *Proceedings of IEEE International Conference on Acoustics, Speech and Signal Processing, 2006. ICASSP 2006*, vol. 2, pp. 933–936.

Y. H. Dandawate and M. A. Joshi, Image compression using generic vector quantizer designed with Kohonen's artificial neural networks. *International Journal of Information*, 3(3): 45–56, 2009.

A. M. Eskicioglu and P. S. Fisher, Image quality measures and their performance. *IEEE Transactions on Communications*, 43(12): 2959–2965, 1995.

W. K. Pratt III, *Digital Image Processing: PIKS Scientific Inside*, Wiley Interscience, New York, 1978.

Z. Wang and A. C. Bovik, *Modern Image Quality Assessment*, Morgan and Claypool Publishers, San Rafael, CA, 2006.

Z. Wang, A. Conrad Bovik, H. Rahim Sheikh, and E. P. Simoncelli, Image quality assessment: From error visibility to structural similarity. *IEEE Transactions on Image Processing*, 13(4): 600–612,, 2004.

Bibliography

I. Avicibas, B. Sankur, and K. Sayood, Statistical evaluation of image quality measures. *Journal of Electronic Imaging*, 11(2): 206–223, April 2002.

D. M. Chandler, Seven challenges in image quality assessment: Past, present, and future research. *ISRN Signal Processing Online Journal*, February 2013, Article ID 905685, 53 pages. http://dx.doi.org/10.1155/2013/905685.

P. Gastaldo, R. Zunino, I. Heynderickx, and E. Vicario, Objective quality assessment of displayed images by using neural networks. *Signal Processing: Image Communication*, 20(7): 643–661, August 2005.

A. Krishna Moorthy and A. Conrad Bovik, A two-step framework for constructing blind image quality indices. *IEEE Signal Processing Letter*, 17(5): 513–516, May 2010.

Q. Li and Z. Wang, General-purpose reduced-reference image quality assessment based on perceptually and statistically motivated image representation. In *Proceedings of the IEEE International Conference Image Processing ICIP*, pp. 1192–1195, October 2008.

Q. Li and Z. Wang, Reduced-reference image quality assessment using divisive normalization-based image representation. In *IEEE Journal of Selected Topics in Signal Processing*, 3(2): 202–211, April 2009.

X. Lv and Z. Wang, Reduced-reference image quality assessment based on perceptual image hashing. In *Proceedings of the IEEE International Conference Image Processing (ICIP)*, pp. 4361–4364, November 2009.

L. Ma, S. Li, F. Zhang, and K. N. Ngan, Reduced-reference image quality assessment using reorganized DCT-based image representation. *IEEE Transactions on Multimedia*, 13(4): 824–829, August 2011.

M. Mrak, S. Grgic, and M. Grgic, Picture quality measures in image compression systems. In *Proceedings of IEEE Region 8, EUROCON 2003. Computer as a Tool*, vol. 1, pp. 233–236, 2003.

Z. M. Parvez Sazzad, Y. Kawayoke, and Y. Horita, No reference image quality assessment for JPEG2000 based on spatial features. *Signal Processing: Image Communication*, 23(4): 257–268, April 2008.

Z. M. Parvez Sazzad, Y. Kawayoke, and Y. Horita, No reference image quality assessment for JPEG2000 based on spatial features. *Signal Processing: Image Communication*, 23(4): 257–268, April 2008.

A. Rehman and Z. Wang, Reduced reference image quality assessment by structural similarity estimation. *IEEE Transactions on Image Processing*, 21(8): 3378–3389, August 2012.

H. R. Sheikh, A. C. Bovik, and L. K. Cormack, No-reference quality assessment using natural scene statistics: JPEG2000. *IEEE Transactions on Image Processing*, 14(11): 1918–1927, November 2005.

A. Shnayderman, A. Gusev, and A. M. Eskicioglu, Multidimensional image quality measure using singular value decomposition. In *Journal of Electronic Imaging 2004*, International Society for Optics and Photonics, Bellingham, WA, pp. 82–92.

E. P. Simoncelli and B. A. Olshanusen, Natural image statistics and neural representation. *Annual Review of Neuroscience*, 24: 1193–1216,, May 2001.

R. Soundararajan and A. C. Bovik, RRED indices: Reduced reference entropic differencing for image quality assessment. *IEEE Transactions on Image Processing*, 21(2): 517–526, February 2012.

S. Suresh, R. Venkatesh Babu, and H. J. Kim, No-reference image quality assessment using modified extreme learning machine classifier. *Applied Soft Computing*, 9(2): 541–552, March 2009.

S. Suthaharan, No-reference visually significant blocking artifact metric for natural scene images. *Signal Processing*, 89(8): 1647–1652, August 2009.

Z. Wang, H. R. Sheikh, and A. C. Bovik, No-reference perceptual quality assessment of JPEG compressed images. In *Proceedings of the IEEE International Conference on Image Processing*, vol. 1, pp. 477–480, September 2002.

Z. Wang and E. P. Simoncelli, Reduced-reference image quality assessment using a wavelet-domain natural image statistic model. *Human Vision and Electronic Imaging*, vol. 5666 of *Proceedings of SPIE*, pp. 149–159, January 2005.

Z. Wang and A. C. Bovik, Reduced and no-reference image quality assessment—The natural scene statistic approach. *IEEE Signal Processing Magazine*, pp. 29–40, November 2011.

Z. Wang, A. C. Bovik, and L. Lu, Why is image quality assessment so difficult? In *Proceedings of IEEE International Conference on Acoustics, Speech, and Signal Processing (ICASSP)*, 4: 3313, 2002.

8

Compressive Sensing

8.1 Introduction

Compressed or compressive sensing (CS) is a mathematical theory of measuring and retaining the *most* important part of the signal while sensing it. It effectively performs dimensionality reduction of a signal in a linear manner. It is one of the most exciting domains of modern times, and there is a deluge of papers and research outcomes available to the researcher. It has opened up new application vistas in the domains of computer science, electrical engineering, applied mathematics, remote sensing, medical imaging, communication, pattern recognition, and many more. Compressive sensing is an interdisciplinary field and draws its power from linear algebra, statistics and random processes, signal processing, optimization, communication theory, and space theory. This chapter is aimed at both the theorists and the practitioners. It will be a review for novice practitioners who would be interested in peeping through the domain, and also act as a quick reference to the theorist. This chapter will focus on the finite dimensional sparse signals and will provide an overview of the basic theory underlying the ideas of compressive sensing. Later in the chapter, we will discuss application based on the compressive sensing theory covered in the former part. We will develop a fragile domain watermarking application using CS. And finally, we present a further line of investigation in this domain: exploiting signal and measurement structure (i.e., use prior knowledge about the signal or physical process to be sensed for further reduction in the sampling rate). For more tangible discussions and simplicity, limited dimensional noncomplex signals are covered in this chapter.

8.2 Motivation for Compressive Sensing

This author has an eternal question in mind: Are the days of analog signal processing over? This question has arisen due to the digital revolution we are going through. Every walk of human as well as animal life has been

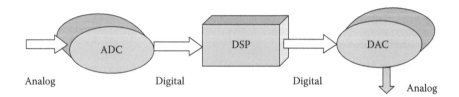

FIGURE 8.1
Analog-to-digital (ADC)–digital signal processing (DSP)–digital-to-analog (DAC) process. The text below the arrows indicates the domain of information processing.

touched by this revolution. Modern digital gadgets used by today's toddlers were only imagined in science fiction half a century ago. But, thankfully the *consumer* of this information to a very large extent remains *Homo sapiens*. Humans consume and deliver analog information (think of a person communicating emotions using numbers instead of a language like Hindi). Thus, a need for analog information and processing is going to prevail until the time when human and machine are alike. At the same time, the digital revolution is improving all walks of life. This essentially means that for getting a *digital advantage*, analog information is converted to the digital domain for processing and then converted back to the analog domain for human consumption. This human-centric conversion process is depicted in Figure 8.1.

The digitization is realized by exploiting the limits of human perception. But it is capable of providing *nearly natural* like signals to the human senses. With the advancement of these technologies, the gap between real signals and its digitally synthesized variants is minimizing. The terms *hi-fi* (high fidelity), *stereophonic, true pictures,* and *HDTV* (high-definition television) are now a part of daily life. Music players synthesize *pure acoustics* and three-dimensional (3D) TVs are capable of producing *real*-like pictures. Billions of bits are buzzing and running around us to make our life simpler and more exciting. As seen in Figure 8.1, the heart of this conversion process is analog-to-digital converters (ADCs) and digital-to-analog converters (DACs) . ADCs form the front end and convert the analog real-world signal into the streams of bits. These bits are then processed in the digital domain using software algorithms running on digital signal processors. ADCs have to deal with a wide range of analog input signals, and they have to sample these input signals. These samples are spaced in time and thus signal information between them is lost. The sampling rate has to be fast enough so that an analog signal can be regenerated from these samples. How fast this sampling rate should be is answered by the fundamental Shannon-Whitakker-Nyquist theorem (Shannon 1949).

Theorem 8.1 (Shannon 1949): An analog signal whose frequency is band limited to B Hz, can be completely recovered from its samples which are spaced at $\frac{1}{2B}$ seconds (secs) apart. ∎

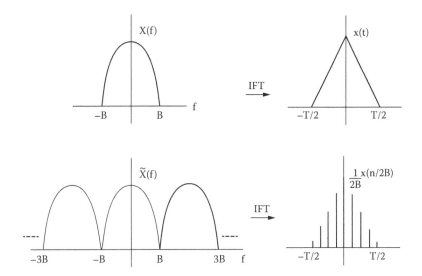

FIGURE 8.2
Relationship between the spectrum of time domain signal and its samples. [IFT: inverse Fourier transform; $X(f)$: frequency domain spectrum; $x(t)$: time domain signal; $\tilde{X}(f)$: frequency spectrum for sampled time domain signal; $x(n)$: sampled time domain signal at a Nyquist rate.]

It is commonly assumed that the signal is band limited (i.e., it has the highest frequency component of B Hz, and it contains no frequencies higher than B Hz). Due to limited frequency variations, these signals can be perfectly recovered from its samples, provided it is sampled at a rate (r) of $2B$, known as the Nyquist rate. This fundamental result is due to Theorem 8.1, which has pioneered the digital revolution. Normally, the $r > 2B$ is required for a perfect reconstruction. The spectrums of time domain signal and its samples exhibit a definite relationship. This relationship is shown in Figure 8.2.

It can be seen from Figure 8.2, that sampling a signal in a time domain produces a periodic sequence in the frequency domain. If the signal is band limited to B, then the bandwidth $[0\ B]$ contains all the frequency components and hence all the information. The original signal can be perfectly recovered from its samples taken at a Nyquist rate. Parseval's theorem links the power and energy spectral densities of analog and digital signals. Therefore, the analog signal can be processed by the DSP.

The increase in wide-band analog signal applications in the 21st century requires exceptionally high sampling rates. This puts a tremendous amount of pressure on the ADCs and the associated back-end devices in the digital chain. ADCs are stressed because they acquire the fast varying signal and gather its samples at a Nyquist rate. Sampling generates a large amount of bits that are subsequently stored and processed. This will further enhance requirements of high storage and tremendous processing power. This becomes a vicious cycle as demands are ever increasing with the advent of

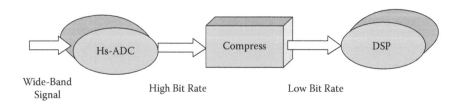

FIGURE 8.3
Wide-band signal acquired by the high-speed ADC (Hs-ADC) and then compressed to remove the perceptually insignificant part of the signal.

new technologies and insistence by innovative applications. One has to pose the question: can this demand be met or at least partially satisfied by the existing hardware? Luckily, limitations of the human sensory systems can be exploited to slightly offset some of these demanding requirements. Most of the acquired data are thrown away without significant perceptual loss. This will relieve subsequent requirements on storage devices and digital processors. This principle has been typified in Figure 8.3.

This arrangement has worked fine until the end of the 20th century but one must note that it has not removed stress from the ADC. The gap between the wide-band application demands and capacity of ADC is widening day by day. Philosophically, so much effort in acquiring the data is wasted as most of the samples are discarded without processing. This philosophy of discarding perceptually insignificant samples is the heart of all the compression algorithms. But new alternatives to a high sampling rate are drawing substantial attention of the research community. The perennial question of just acquiring the part of the signal to be retained was first addressed by Donoho (2006). He suggested merging the sampling with compression for (1) avoiding wastage of samples and (2) relieving ADCs of the relentless pressure of acquiring wide-band signals at a rate greater than the Nyquist rate. To the best of this author's knowledge, surprisingly, not many researchers have posed this hypothesis from the late 1940s to early 2000. The idea of simultaneously compressing and sampling is shown in Figure 8.4. The dashed block encompasses both the functions seamlessly and the output is a CS measurement, which has lower dimensionality in comparison to the original analog signal.

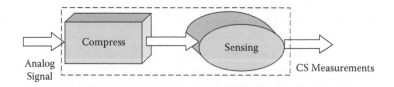

FIGURE 8.4
Idea of simultaneous compression and sampling. (From Donoho, *IEEE Trans. Inf. Theory,* 52(4):1289–1306, September 2006.)

8.3 Basics of Simultaneous Compression and Sensing

CS is a skeleton for acquiring the signals that are sparse with extremely low sampling rates. The number of samples required to represent the signal reduces considerably if the sparse signal to be acquired can be represented with the known basis. This is the fundamental driving force for compressed sensing. CS is a mechanism that directly senses the imperative information from the sparse signal at a low sampling. The current developments in CS are due to the fundamental work done by Donoho (2006), Candes and Tao (2006), Candes, Romberg, and Tao (2006), and Baraniuk (2007). They showed that finite dimensional signal can be almost exactly recovered from its linear measurements. These measurements are nonadaptive in nature and have lower dimensionality as compared to the finite dimensional signal under consideration. Even though initial work focused on the finite dimensional signal recovery, it was a paradigm shift in the strategy of sampling the signal. The exact recovery of the sparse signal from its linear measurements, which has lower dimensionality as compared to a signal, has resulted in the nomenclature of the *compressed sensing*.

Can you find a common link between the conventional wisdom of compression and the new thinking of the compressed sensing? If we observe carefully, both schools of thought are exploiting *sparsity* as a signal structure. In the conventional compression, one of the most popular techniques is *transform coding*. It typically applies a frequency transformation on the signal and represents it in terms of the fixed basis function $\{\varnothing_i\}_{i=1}^{N}$ for $\mathbb{R}^N$. Signal $x \in \mathbb{R}^N$ is expressed by N coefficients

$$\{\theta_i\}_{i=1}^{N} \quad \text{for} \quad \mathbb{R}^N \quad \text{i.e.} \quad x = \sum_{i=1}^{N} \varnothing_i \theta_i \tag{8.1}$$

Succinctly, in the matrix form $x = \Phi\theta$ with $\theta \in \mathbb{R}^N$, where Φ is an $N \times N$ matrix with $\varnothing_i$ forming columns of a matrix and θ forming $N \times 1$ coefficient vector. Sparsity implies that the signal with length N can be represented using k nonzero coefficients with $k \ll N$. Such a signal is known as a k-sparse signal. The indices of such nonzero coefficients are called *support* of θ, denoted as $S(\theta)$. The *k-sparse* signal can be very well approximated by the largest k nonzero coefficients such that $x \cong \Phi\theta$.

Such a signal can be approximately reconstructed if we know the location and magnitude of the k largest nonzero coefficients. This forms the foundation for the transform coding and resulted in popular compression standards like JPEG, JPEG 2000, MPEG, and MP3. If one carefully notes, most of the natural images possess a strong correlation in the neighborhood; therefore, these signals can be sparsified by de-correlating them in the transform domain. Such a signal is efficiently represented using a wavelet transform. Figure 8.5 shows an example of wavelet domain representation for such an image.

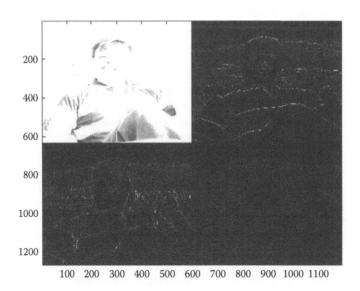

FIGURE 8.5
Wavelet domain representation of the natural image indicating approximate, horizontal, vertical, and diagonal details in an image.

Most of the coefficients in the finer scales of the wavelet domain have extremely small values. Such coefficients can be rounded to zero or thresholding can be applied to them to obtain the k-sparse representation (i.e., the signal will have only k nonzero coefficients). When measuring the fidelity error during signal reconstruction in terms of least square error (LSE) sense, such a process of thresholding coefficients results in the best approximation. Such a signal is represented in terms of k basis functions. Figure 8.6 represents an original image and its k coefficients approximation reconstructed version. In this case, 20% of the largest coefficients are retained during reconstruction.

It is also interesting to compare the conventional sampling theory to CS-based sampling. Table 8.1 shows the comparison between the two approaches.

FIGURE 8.6
Original image and its sparse approximation retaining 20% of the largest coefficients. The coarser details are retained with some loss of finer details.

TABLE 8.1

Comparison among the Sampling Strategies: Conventional and Compressive Sensing Based Samplings

Whittaker-Shannon-Kotelnikov (WSK) Sampling	Compressed Sensing Based Sampling
The sampling rate is greater than twice the maximum signal frequency in the band-limited signal.	It follows the sub-Nyquist sampling rate. Reducing the sampling rate is one of the main motivating factors for development of CS.
Samples are either evenly spaced in time (uniform sampling) or they are unevenly spaced in time (nonuniform sampling).	The samples are taken using a randomized measurement matrix.
Nonuniform sampling is based on Lagrange interpolation, and it is a generalized form of uniform sampling.	The samples (measurements) are an inner product between sparse signal and specialized measurement matrix.
Sampling theory was developed for continuous time signal with infinite dimensions.	Initial work on CS focused on finite dimensional signal in $\mathbb{R}^N$. However, the current focus is developing CS for continuous time, infinite dimensional signal (Duarte and Eldar 2011; Eldar and Kutyniok (2012).
No underlying signal structure is assumed during sampling.	Initially CS theory was developed for signals that exhibit sparsity. Now the research community is looking for representation of a signal structure beyond sparsity.
As a consequence of high sampling rate, a huge chunk of data is generated. This can be tackled by applying the transform coding–based compression. A very high value of compression degrades signal quality.	If a signal structure over and above sparsity is applied to an analog signal, it results in a high degree of compression without significant degradation in signal quality.
It does not consider the hardware implementation of the real-world applications. The underlying physics of the signal except the band limitation is not exploited.	The current CS research is driven by the real-world hardware implementation. It considers the signal structure over and above sparsity and also considers structured nonrandom measurement matrix.
The sampling functions are deterministic, and the sampling mechanism is nonlinear.	The randomness plays a very important role in the design of the measurement matrix for taking the linear measurements.
Ideally the signal reconstruction involves the use of sinc functions.	The signal recovery from the linear measurements involves use of convex optimization techniques and linear programming.
Scaled sinc function is placed at each sample location, and then an interpolation is performed on these functions to recover an original signal. This results in an interpolation error because one cannot produce an ideal sinc pulse as it can only be approximated.	The signal recovery is done through complex nonlinear methods (Eldar and Kutyniok 2012). These methods have been shown to work well in the presence of noise.
It can be summarized as nonlinear sampling mechanism with linear recovery, which is intuitive to understand.	It can be summarized as linear measurements with nonlinear recovery methods that are difficult to understand.

8.4 Basic Compressive Sensing Framework

If the original signal is sparse, then the CS framework tries to recover signal x, which is a $N \times 1$ vector from its measurement y, which is a $M \times 1$ vector with $M \ll N$. The samples are captured as follows if the signal is sparse:

$$y = \Psi x \tag{8.2}$$

where $\Psi \in \mathbb{R}^{M \times N}$ is a random measurement matrix. It is one of the fundamental building blocks of the CS system. In case the original signal is nonsparse, then it can be sparsified by using a *sparsifying dictionary* of basis $\{\varnothing_i\}_{i=1}^{N}$, and measurements y are obtained as follows:

$$y = \Psi \Phi \Theta \tag{8.3}$$

where $y \in \mathbb{R}^M, \Psi \in \mathbb{R}^{M \times N}, \Phi \in \mathbb{R}^{N \times N}, \Theta \in \mathbb{R}^N$.

The goal of CS would be to recover $x \in \mathbb{R}^N$ with the known measurements $y \in \mathbb{R}^M$ and Ψ. We have to find the signal from the given class $\mathbb{C}_k$ such that $y = \Psi x$ matches. Class $\mathbb{C}_k$ is the collection of all possible k-sparse signals. Here $M \ll N$, and the goal is to push M as close as possible to k. During recovery of the sparse signals, CS makes an exhaustive search for x, which has the fewest nonzero entries in class $\mathbb{C}_k$, which can be solved as an ℓ_0 optimization problem:

$$\hat{x} = \arg\min_{x \in \mathbb{R}^N} \|x\|_0 \text{ such that } y = \Psi x \tag{8.4}$$

$\|\cdot\|_0$ represents ℓ_0 norm of a vector that represents the number of nonzero components in it. This norm is difficult to track due to the presence of zero in its root and power. The definition of ℓ_0 norm as used for the practical purpose is

Definition 8.1:

$$\|x\|_0 = \#(i | x_i \neq 0)$$

that is, total number of nonzero elements in the signal. ∎

However, ℓ_0 norm is nonconvex (Candes, Romberg, and Tao 2006; Natarajan 1995), and it is known that nonconvex optimization problems are difficult to solve Vavasis (1991). It has also been observed that ℓ_0 optimization is computationally very expensive: it is a nondeterministic polynomial-time (NP) hard problem. As an example, consider the following computations for illustrating the computational complexity. Assume that $\Psi \in \mathbb{R}^{M \times N}$ with $M = 1000$ and $N = 5000$ and assume that it is a 100-sparse ($k = 100$) signal. We have to extensively search for appropriate 100 columns among 5000, meaning a search of $\binom{5000}{100}$, practically an infinite number. For any general matrix Ψ, even finding an approximate solution to the true minima is also NP hard. The complexity in solving nonconvex optimization arises due to use of nonconvex energy function formulation. Thus, it is mandatory to look at some of the interesting questions arising due to this situation: (1) Can this problem be solved by any other mechanisms? (2) How can we get an estimated solution and what level of estimation accuracy is acceptable? (After all, engineers always look for the workable approximate solutions.) (3) What kind of approximation will yield the solution closer to the desired one? In this chapter we will deal with these questions, although succinctly. Readers may refer to many such questions that have been handled in the literature (Elad 2010).

For the sake of understanding the bigger picture, let us review the definition of a convex set.

Definition 8.2: A set Ω is convex if $\forall x_1, x_2 \in \Omega$ and $\forall t \in [0,1]$ the convex combination $= tx_1 + (1-t)x_2 \in \Omega$. ∎

As can be seen from Figure 8.7a that an object is convex in Euclidean space if a line segment joining two points inside the set also falls within the set (i.e., every point on the straight line falls within the set). Figure 8.7b indicates the nonconvex set. One of the ways to handle an intractable nonconvex optimization problem is to solve a similar *convex* optimization problem, which is known as *convex relaxation* (Candes and Tao 2009). However, one must not conclude that there is no solution to a nonconvex problem, but be aware of the consequences arising due to nonconvexity. For convex optimization problems several optimization techniques are available (Boyd and Vandenberghe 2004). Candès and Tao (2009) proposed to use convex ℓ_1

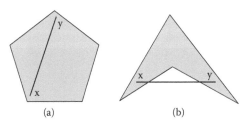

(a)   (b)

FIGURE 8.7
(a) Convex set. (b) Nonconvex set.

norm as an *approximation* to the nonconvex ℓ_0 norm optimization problem. Specifically,

$$\hat{x} = \arg\min_{x \in \mathbb{R}^N} \|x\|_1 \text{ such that } y = \Psi x \tag{8.5}$$

where $\|\cdot\|_1$ represents ℓ_1 norm of a vector defined as

Definition 8.3:

$$\|x\|_1 = \sum_i |x_i|$$

∎

The set $\Omega = \{x | y = \Psi x\}$ is convex; thus, the optimization problem posed above is also convex. A strictly posed convex problem leads to a close form solution and is guaranteed to converge at the local minima (Bruckstein, Donoho, and Elad 2009). Candes and Tao (2009) proposed to find all the values of x_i's for which ℓ_1 norm is as small as possible. The $\|\cdot\|_1$ is not strictly convex, as it may have more than one solution and in the worst case, an infinite number of solutions. The problem also requires a good computing power as it has to flounce through every possibility in the solution. It means that the algorithm has to search through an infinite number of possible solutions. However, these solutions include: (1) clustered around in a convex set, as all optimal solutions will have an ℓ_1 penalty and their combination would also be convex; (2) the set is bounded; (3) among them at least one has at most k nonzero elements. It has been shown in the literature (Candes, Romberg, and Tao 2006; Chen, Donoho, and Saunders 1998; Olshausen and Field 1996) that under certain constraints, ℓ_1 norm optimization provides a very close solution to the ℓ_0 norm problem. It tends to have a sparse solution. However, it may not always be feasible to work under the specified constraints, and it results in a solution that may not closely approximate the actual one. However, it is still the best possible solution as compared to other approaches. It can be seen from Figure 8.8, that ℓ_1 norm solution approximates the ℓ_p $\forall p < 1$ norm

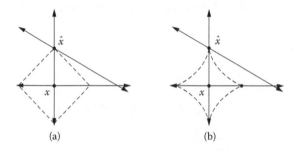

(a)   (b)

FIGURE 8.8
(a) ℓ_1 norm approximation of a point x in $\mathbb{R}^2$. (b) ℓ_p norm with $p < 1$ approximation of a point x in $\mathbb{R}^2$.

solution for a sparse signal. Figure 8.8 indicates the best approximation of a point in $\mathbb{R}^2$ using the ℓ_p norms with $p < 1$.

8.5 Heart of Compressed Sensing: Measurement Matrix

One of the most important design criteria for a measurement matrix is to have a one-to-one linkage between measurements and a signal. The linear measurements $y = \Psi x$, where $\Psi \in \mathbb{R}^{M \times N}$ is a random measurement matrix, are static in a sense that columns $\{\Psi_i\}_{i=1}^N$ are fixed *a priori*. They do not depend on the signal or on the prior measurements. The most fundamental questions in the design of a measurement matrix are as follows: (1) How much of the information about signal x is retained in its linear measurements y? (2) How can the linear measurements y uniquely represent x? (3) How can the original signal be recovered from its measurement? In case the signal is k-sparse with $\Psi \in \mathbb{R}^{M \times N}$ and M $\ll$ N, we will see later that using an appropriate measurement matrix, a variety of algorithms are available for exact recovery. The measurement matrix Ψ is designed to reduce the dimensionality, but it results in a rank-deficient matrix. It means that the information content of the matrix is low, and based on the dimension theorem of a vector space, it results in a nonempty null space $\mathcal{N}(\Psi)$ defined as follows.

Definition 8.4: $\mathcal{N}(\Psi) = \{x \colon \Psi x = 0\}$. ∎

This means that the matrix has an infinite number of solutions, and this further implies that an infinite number of signals $x \in \mathbb{R}^N$ will result in the same measurement vector (i.e., $y = \Psi x_1 = \Psi x_2 = \Psi x_3 = \ldots$) for a given measurement matrix Ψ. This will not allow recovery of all of the sparse signals using their measurements. It is clear that the design of the CS measurement matrix should resolve this ambiguity. It should be designed in such a manner that for distinct k-sparse signals $x_1, x_2, \ldots, x_n \in \mathbb{C}_k$ must have unique measurements $y_1, y_2, \ldots, y_n$ (i.e., $\Psi x_1 \neq \Psi x_2 \neq \Psi x_3 \neq \ldots \Psi x_n$) even though $M \ll N$. We call this a *uniqueness property* of Ψ. This is possible by considering that the signal belongs to a sparse class while designing Ψ.

With $M > k$, the k-sparse signal for support $S(x)$ will result in k unknown. In this case, columns of Ψ can be pruned to include only those that correspond to indices of $S(x)$. Assuming that the resulting matrix $\Psi_{S(x)}$ is a full-column rank, one can obtain the k-sparse signal using Moore-Penrose (pseudo inverse) as follows:

$$\hat{x}_{S(x)} = \left(\Psi_{S(x)}{}^T \Psi_{S(x)} \right)^{-1} \Psi_{S(x)}{}^T \tag{8.6}$$

where $(\cdot)^T$ indicates the matrix transpose, $(\cdot)^{-1}$ indicates the matrix inverse. But a full-column rank matrix may not always be realizable, so let us look at the important properties of Ψ which characterize its *uniqueness*. One of them is the *spark* (Donoho and Elad 2003) of a matrix. The spark is derived from the words "sparse" and "rank."

8.6 Important Properties of the Measurement Matrix

8.6.1 Spark

Theorem 8.2 (Donoho and Elad 2003): If *spark* $(\Psi) > 2k$, then measurement $y \in \mathbb{R}^M$ (i.e., $y = \Psi x$) uniquely represents a k-sparse signal belonging to class $\mathbb{C}_k$. This result demonstrates that larger spark can sustain signals with higher sparsity level.   ∎

Definition 8.5: The spark of a given matrix Ψ is the *smallest* number of columns in it that are linearly dependent—that is,

$$spark(\Psi) = \min_{x \neq 0} \|x\|_0 \text{ such that } \Psi x = 0.$$ ∎

The spark of the measurement matrix is used to ensure stability and consistency. Knowing the value of spark can be very revealing: the larger its value, the better it gets.

Some of the important properties of spark are

$$spark\ (\Psi) \in \{1, 2, \ldots, n\} \cup \{+\infty\}$$

$$spark\ (\Psi) = +\infty \leftrightarrow rank\ (\Psi) = n$$

$$spark\ (\Psi) = 1 \text{ iff } \Psi \text{ has a zero column}$$

$$if\ spark\ (\Psi) \neq +\infty, \text{ then } spark\ (\Psi) \leq rank\ (\Psi) + 1$$

When dealing with approximate sparse signals, one must ensure that null space $\mathcal{N}(\Psi)$ does not contain any compressible signal in addition to the sparse vectors (Cohen, Dahmen, and DeVore 2009). It has also been shown (Eldar and Kutyniok 2012) that *spark* $(\Psi) \in [2, M+1]$ constrains the requirement that dimensionality of measurement M be greater than $2k$. If Ψ is derived from independent and identically distributed (i.i.d.) Gaussian distribution, then *spark* $(\Psi) = M + 1$ with probability close to one. It has been shown that computing the spark of a matrix is a NP hard problem (Tillmann and Pfetsch 2013). It has a huge conjunctional computational difficulty—that is, search over all $\begin{pmatrix} n \\ k \end{pmatrix}$ submatrices of a full spark matrix. Thus, we have to look for another property of Ψ which guarantees uniqueness but with lower computational load. The coherence of a matrix is such a property.

8.6.2 Coherence

The concept of coherence was introduced by Donoho and Huo (2001), and it is used widely in the domain of CS. Coherence is used with several basis pursuit algorithms to correctly identify the sparse signal. Mutual coherence is computationally cheaper as compared to computationally exorbitant spark. The concept of mutual coherence is based on the generalization of the Gram-Schmidt matrix:

$$\mho^T \mho = \begin{bmatrix} I & \Psi^T \Phi \\ \Phi^T \Psi & I \end{bmatrix} \tag{8.7}$$

It is the largest entry in $\mho^T \mho$ barring the diagonal values. The definition is as follows:

Definition 8.6 (Donoho and Elad 2003; Donoho and Huo 2001; Tropp, 2006; Fuchs 2004): The mutual coherence of a matrix Ψ is the biggest absolute value of the cross-correlations between its columns—that is,

$$\Gamma(\Psi) = \max_{1 \le i \ne j = N} \left| \psi_i^T \psi_j \right|$$

where ψ_i's are the columns of matrix Ψ, and they are normalized (i.e., $\psi_i^T \psi_i = 1$). The coherence is lower bounded by the Welch bound (Duarte and Eldar 2011; Eldar and Kutyniok 2012; Welch 1974)

$$\Gamma(\Psi) \in \left[\sqrt{\frac{N-M}{M(N-1)}}, 1 \right]$$

provided Γ is a full-rank matrix. Normally, $N \gg M$, in which case the Welch bound reduces to

$$\Gamma(\Psi) = \sqrt{\frac{1}{M}}$$

This is one way to describe the dependency between two columns of Ψ. For unitary matrices, this value will be zero. In this case, with $N \gg M$, mutual coherence will be positive and its value should be as small as possible, which characterizes its desirable property. It has also been shown (Donoho and Elad 2003; Elad 2012; Gribnoval and Nielsen 2003; Tropp 2004) that for each measurement vector $y \in \mathbb{R}^M$ and for $y = \Psi x$ there exist a signal $x \in \mathbb{C}_k$, if the following is satisfied,

$$k < \frac{1}{2}\left(1 + \frac{1}{\Gamma(\Psi)}\right)$$

This result in conjunction with the Welch bound gives the supremum on sparsity k for the coherence (Duarte and Eldar 2011), which is $\mathcal{O}(\sqrt{M})$. ■

The relationship between spark and mutual coherence is as follows (Duarte and Eldar 2011; Eldar and Kutyniok 2012; Donoho and Elad 2003; Elad 2010):

$$spark \ (\Psi) \geq 1 + \frac{1}{\Gamma(\Psi)} \tag{8.8}$$

It is also important to quantify the performance of the CS recovery algorithm when dealing with the signal that is partially sparse or a nonsparse signal. Let $\mathbb{Q} : \mathbb{R}^M \to \mathbb{R}^N$ be one recovery method. Then the following guarantee will ensure exact recovery of the k-sparse signal (Eldar and Kutyniok 2012) $\forall x$:

$$\|\mathbb{Q}(\Psi x) - x\|_2 \leq \rho \frac{\min_{\hat{x}} \|x - \hat{x}\|_1}{\sqrt{k}}, \text{ where, constant } \rho > 0 \tag{8.9}$$

This guarantee also quantifies the robustness of recovery in the case of a nonsparse signal. It will depend on how well the k-sparse signal approximates the original signal. This type of guarantee can be established by studying the null space property (NSP) (Donoho and Elad 2003; Elad 2010; Chen, Wang, and Wang 2013) of a measurement matrix Ψ. It is formally defined in the next section.

8.6.3 Null Space Property

Definition 8.7: A matrix $\Psi \in \mathbb{R}^{M \times N}$ satisfies null space property in relation to a set $\Lambda \subset \{1, 2, \ldots, n\}$, if $\|h_\Lambda\|_1 < \|h_{\Lambda^c}\|_1 \ \forall h \in \mathcal{N}(\Psi) \setminus \{0\}$, where h is a vector, and $\mathcal{N}(\Psi)$ is a null space of matrix Ψ. ■

NSP conceptualizes that vectors in $\mathcal{N}(\Psi)$ should not be clustered on a limited set of indices. If NSP is satisfied, then $h = 0$ is the only k-*sparse* vector in the null space of Ψ. It has also been shown (Eldar and Kutyniok 2012; Chen, Wang, and Wang 2013) that $2k$-NSP ascertain the assurance for ℓ_1 norm-based sparse recovery. NSP is a necessary and sufficient condition for unique recovery. Authors have shown (Eldar and Kutyniok 2012) that NSP is pertinent only for the noiseless situation; however, it was shown (Chen, Wang, and Wang 2013; Aldroubi, Chen, and Powell 2012; Sun 2011) that it is applicable for compressible signals in the noisy environment as well. However, stability analysis for the NSP cannot be easily generalized.

8.7 Uniqueness Guarantee in the Presence of a Noise

Properties like spark, coherence, and to some extent NSP guarantee the *uniqueness* of mapping between signal and its measurement. However, it would be interesting to look at this issue in the presence of a noise (Chi et al. 2011). The noise can creep into the system during the sensing stage while acquiring a signal, affecting the measurements. Such a noise is usually modeled as an additive noise:

$$y = \Psi x + n \tag{8.10}$$

This will result in inaccurate measurements. In CS, it is assumed that Ψ is available during sparse signal recovery. The modified $\Psi' = \Psi + \Delta$ will cause changes during the signal recovery. It is not possible to guarantee the uniqueness under the noisy conditions, but it is advantageous to have a measurement process robust to such a noise. If one can manage that distance between two measurement vectors is comparative to the distance between two signals in $\mathbb{R}^N$, then the sparse recovery in the presence of noise is also feasible. More specifically, $\mathcal{D}(y, y') \propto \mathcal{D}(x, x')$, where $\mathcal{D}(\cdot, \cdot)$ is the distance metric, (y, y') represents measurements, and (x, x') represents the sparse signals. This property give an interesting result: if the $\mathcal{D}(x, x')$ is very large in a high-dimensional space, then signal corruption due to noise n will not lead to the same measurement vectors (i.e., $y \neq y'$). This property yielding uniqueness is captured by *restricted isometric property* (RIP).

8.7.1 Restricted Isometric Property

Let us review the concepts of *bases* and *frames* to delve into the understanding of RIP. $\{\emptyset_i\}_{i=1}^N$ forms the basis for $\mathbb{R}^N$ if (1) they span it, meaning every vector in $\mathbb{R}^N$ is represented by the linear combination of these basis vectors or (2) they are linearly independent. The signal can be represented in terms of this basis as $x = \sum_{i=1}^N \emptyset_i \theta_i$, where $\{\theta_i\}_{i=1}^N$ forms a unique coefficient of the signal. The concept of basis can be generalized to incorporate the notion of linearly dependent vectors, and it is known as a *frame* (Casazza and Kutyniok 2012).

Definition 8.8: A frame is a set of vectors $\{\emptyset_i\}_{i=1}^V$ in matrix $\forall \in \mathbb{R}^{U \times V}$ with $U < V$ such that for $\forall x \in \mathbb{R}^U$ the following is true, if there exist constants α, β such that $\alpha \|x\|_2^2 \leq \|\Phi^T x\|_2^2 \leq \beta \|x\|_2^2$, where, $0 < \alpha \leq \beta < \infty$. ∎

The linearly independent columns will exist due to $\alpha > 0$, and this also ensures that $\Phi \Phi^T$ is invertible. With $\Phi \in \mathbb{R}^{U \times V}$, α, β corresponds to the smallest and largest eigenvalues of $\Phi \Phi^T$, correspondingly. Frames provide more elaborative representations for the sparse signal (Bodmann,

Casazza, and Kutyniok 2011; Boufounos, Rauhut, and Kutyniok 2011) due to redundancy (existence of linearly dependent vectors). One simple way to obtain the possible set of coefficients is through the Moore-Penrose inverse as follows $\Theta = \Phi^T (\Phi \Phi^T)^{-1} x$. This is an optimal set of coefficients in terms of ℓ_2 error norm.

Mathematically, isometry is a function between the two spaces, which has a property to preserve the distance between each pair of points. This mapping is one to one from one between metric spaces. The restricted isometry property provides stronger conditions for establishing uniqueness in the presence of noise due to quantization effects or contamination while measuring the signal. It was introduced by Candes and Tao (2005) as a condition on the measurement matrix.

Definition 8.9: A matrix Ψ satisfies (k, δ), restricted isometry property of order k if for $\forall x \in \mathbb{C}_k \ \exists \ \delta \in (0,1)$ such that $(1-\delta)\|x\|_2^2 \leq \|\Psi x\|_2^2 \leq (1+\delta)\|x\|_2^2$. ∎

Due to the k-RIP, the submatrices of Ψ with size $M \times k$ will be distance-preserving ones and thus close to being isometric. If Ψ has $2k$-RIP, then by definition of RIP it can be seen that the measurement matrix will preserve the distance between any two k-sparse vectors (Eldar and Kutyniok 2012). RIP of order $2k$ $(2k, \delta_{2k})$ means that all sets of $2k$ columns of Ψ are linearly independent, that means $spark(\Psi) > 2k$. Also one may note that if matrix Ψ satisfies (k, δ_k)-RIP, then for any $k' < k$, Ψ automatically satisfies $(k', \delta_{k'})$ RIP with $\delta_{k'} < \delta_k$ (Needell and Tropp 2009). It has also been shown (Davenport 2010) that Ψ satisfying the restricted isometry property also satisfies the null-space property. This shows that RIP is stronger than NSP. The following theorem links the two.

Theorem 8.3 (Davenport 2010): If Ψ satisfies the $(2k, \delta_{2k})$ RIP with $\delta_{2k} < \sqrt{2} - 1$, then Ψ satisfies NSP of order $2k$ with constant $\mathfrak{C} = \frac{2}{1-(1+\sqrt{2})} \frac{1}{\delta_{2k}}$. ∎

As we have seen earlier, to validate RIP for a measurement matrix will require a combinatorial search over $\binom{n}{k}$ submatrices. Therefore, one may link RIP to the coherence for reducing the search space (Cai, Xu, and Zhang 2009).

Lemma (Cai, Xu, and Zhang 2009): If measurement matrix Ψ is normalized to unit norm and coherence $\Gamma = \Gamma(\Psi)$, then Ψ has a (k, δ_k) RIP with $\delta_k \leq (k-1)\Psi$. ∎

Now having defined the pertinent properties of the measurement matrix, let us discuss the type of matrix that satisfies them.

8.8 Building a Sensing Matrix

In this section, we define the specific matrices that suit the CS construction. A $M \times N$ *Vandermonde* matrix V constructed using N scalars has *spark* $(V) = M + 1$ (Cohen, Dahmen, and DeVore 2009) and $\mathfrak{V}_{i,j} = a_i^{j-1} \forall \ i, j$ (i.e., it has geometric progression in each row). Matrix V suffers from poor conditioning as the values of $a_i'^s$ grow with the increase in the value of N. It is measured by the condition number $\mathcal{C}$ of the linear system of equations $Ax = b$, which exhibit the following relationship:

$$\frac{\Delta x}{x + \Delta x} \leq \|A\| \|A^{-1}\| \frac{\Delta b}{b} \text{ and } \frac{\Delta x}{x} \leq \|A\| \|A^{-1}\| \frac{\Delta A}{A} \tag{8.11}$$

This relationship shows that changes on the RHS are amplified by a factor of $\mathcal{C} = \|A\| \|A^{-1}\|$, and it represents the condition number of the matrix. It is related to the finite precision of the machine, and V is poorly conditioned and will bring instability into the CS system. Formally, for the Vandermonde matrix $\mathcal{C}_\infty = \|\mathfrak{V}\|_\infty \|\mathfrak{V}^{-1}\|_\infty$, where $\|\cdot\|_\infty$ represents ℓ_∞ norm.

Definition 8.10: $\|\mathfrak{V}\|_\infty = max\{N, \sum_{\mu=1}^{N} |x_\mu|^{N-1}\}$ defines the infinity norm of the Vandermonde matrix. ∎

Another type of matrix that satisfies the conditions required for uniqueness is the Gabor frame. These are seen as a collection of frequency and time shifts of a nonzero vector in the Hilbert space $\mathbb{H}^n$. Gabor frames are popular due to the following reasons (Bajwa, Calderbank, and Jafarpour 2010): (1) The n numbers describing the seed can completely specify the Gabor frame, (2) the multiplication of Gabor frames can be handled efficiently by the fast Fourier transform (FFT), and (3) many applications in communication and image processing showcase Gabor frames. The process for Gabor frame generation is as follows: specifically, let $\hbar \in \mathbb{H}^n$ be a unit norm seed vector and let $G \in \mathbb{H}^n$ be the $n \times n$ time shift matrix generated as follows:

$$G(\hbar) = \begin{bmatrix} \hbar_1 & \hbar_n & \cdots & \cdots & \hbar_2 \\ \hbar_2 & \hbar_1 & \ddots & \ddots & \vdots \\ \vdots & \vdots & \ddots & \ddots & \vdots \\ \vdots & \vdots & \ddots & \ddots & \hbar_n \\ \hbar_n & \hbar_{n-1} & \cdots & \cdots & \hbar_1 \end{bmatrix} \tag{8.12}$$

Now generate the discrete sinusoid with frequency

$$\omega_m = \left[e^{j2\pi 0\frac{m}{n}}, \ldots, e^{j2\pi(n-1)\frac{m}{n}} \right]^T, \quad m \in \{0, 1, \ldots, n-1\}$$

Generate the diagonal matrix $\mathfrak{D}_m = diag(\omega_m)$. Then the Gabor frame is a $\mathcal{G} = [\mathfrak{D}_0 G, \mathfrak{D}_1 G, \ldots, \mathfrak{D}_{n-1} G]$ is a $n \times n^2$ block matrix. The columns of G are given the downward circular shifts. It has been shown that $\|\mathcal{G}\|_2 = \sqrt{n}$. Average coherence of Gabor frame is bounded as follows (Bajwa, Calderbank, and Jafarpour 2010):

$$\Gamma(\mathcal{G}) \leq \frac{n\hbar_{max}\left(\sqrt{n} - \hbar_{min}\right) + 1 - n\hbar_{min}^2}{n^2 - 1} \tag{8.13}$$

where $\hbar_{max} = \max_i |\hbar_i|$ and $\hbar_{min} = \min_i |\hbar_i|$. It states that the average coherence of Gabor frame G cannot be too large. In terms of sparse signal recovery, this is the best possible bound on G, nothing better can be expected than it. Measurement matrices with the smallest possible coherence are desirable in the sparse signal recovery. The Gabor frame generated from sequence

$$\left\{ \frac{1}{\sqrt{n}} e^{j2\pi\frac{\tau^3}{n}} \right\}_{\tau=0}^{n-1}$$

will have the smallest worst-case coherence. The sequence is known as an *Alltop* sequence (Alltop 1980). The autocorrelation function for this sequence dies out very quickly. Moreover, the autocorrelation coherence for a Gabor frame generated using the Alltop sequence will be

$$\Gamma(\mathcal{G}) = \max_{i,j:i\neq j} |\langle x_i, x_j \rangle| \leq \sqrt{\frac{1}{n}} \tag{8.14}$$

However, this matrix requires $M = \mathcal{O}(k^2 \log(N))$ number of measurements to recover the k-sparse signal. This restricts the use of Gabor frames as measurement matrices for large values of N and k.

The good thing is that randomness again comes to the rescue and these difficulties imposed by the measurement matrix formulation using Vandermonde and Gabor frames can be overcome. This is one of the foremost surprising aspects of CS that a random measurement matrix with entries drawn for appropriate distribution will provide uniqueness and compression. This fact is highlighted by the Johnson-Lindenstrauss (JL) lemma (Baraniuk et al. 2008), which can be seen as working with relative distances among the points and it does not change with scaling.

JL Lemma (Baraniuk et al. 2010): It articulates that a small set of points in a very high-dimensional space can be mapped to a lower dimensional space without significant change of distance among them.   ∎

The Bernoulli, Gaussian, or sub-Gaussian satisfies RIP with a high probability and allows for sparse signal recovery. The Bernoulli matrix takes an equally probable $\pm\frac{1}{\sqrt{M}}$, and Gaussian matrix columns are independent, following normal distribution with $\mu = 0$ and $\sigma = 1/M$. The result is showcased via the following theorem.

Theorem 8.4 (Baraniuk et al. 2008): Let $\Psi \in \mathbb{R}^{M \times N}$ be a Gaussian or Bernoulli random matrix. Let ϵ & $\rho \in (0,1)$. Assume $M \geq c\rho^{-2}(k\ln(\frac{N}{k})+ln(\epsilon^{-1}))$ for a constant $c > 0$. Then with $1 - \epsilon$ probability, the restricted isometric constant of Ψ satisfies $\rho_s \leq \rho$. ∎

All k-sparse vectors with L1 minimization are obtained if M satisfies the above requirement. It also predicts the stable and robust recovery. The random matrices with i.i.d. distributed members drawn from a continuous distribution will have a *spark* $(\Psi) = M + 1$ with a high probability. The Gaussian distributed measurement matrix will have high (k, δ)-RIP provided M satisfies the following (Baraniuk et al. 2008):

$$M = \mathcal{O}\left(\frac{klog\left(\frac{N}{k}\right)}{\delta^2}\right) \tag{8.15}$$

As a passing note, one must realize that the design of the measurement matrix is often application dependent and subjected to the physical constraints of the application. Therefore, it is desirable to explore the structure in the sensing architecture to create the *structured* measurement matrix. These matrices may not be completely random and showcase the structure inherited from real-world applications. Interested readers may consult the literature (Duarte and Eldar 2011) for an excellent review on the matter.

8.9 Compressive Sensing Sparse Recovery

CS recovery has become almost identical to the ℓ_1 minimization. There are a number of well-known algorithmic approaches for sparse signal recovery from its measurement in CS apart from ℓ_1 minimization. However, some of the approaches exploiting sparsity as a signal structure are older than CS and can be employed in CS recovery (Tropp and Wright 2010). Figure 8.9 indicates the popular approaches in CS recovery algorithms. The choice of algorithm selection for a sparse recovery is based on various factors, namely, (1) signal reconstruction timing from the measurement vector; (2) number of measurement required for recovery to determine the storage requirements; (3) the simplicity of the implementation; (4) possible portability to the hardware for execution; and (5) fidelity of the signal recovery. Even though the ℓ_1 minimization remains the popular choice, we also cover alternative algorithms for CS recovery. We will compare these algorithms on some of the factors mentioned above.

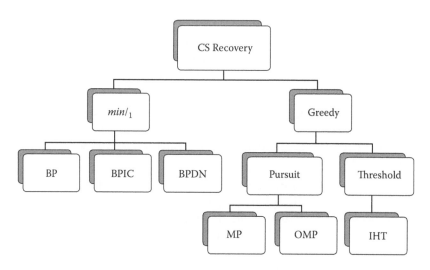

FIGURE 8.9
Various CS recovery algorithms: *min l₁* : ℓ_1 minimization. (BP: basis pursuit; BPIC: basis pursuit with inequality constraint; BPDN: basis pursuit denoising; MP: matching pursuit; OMP: orthogonal matching pursuit; IHT: iterative hard thresholding.)

8.9.1 ℓ_1 Minimization Framework

This is one of the most popular frameworks in sparse signal recovery. Its power stems from the fact that it converts the search into a convex problem and provides accurate recovery. The basic noiseless recovery formulation is defined as follows (cf. Equation 8.5):

$$\hat{x} = \arg \min_{x \in \mathbb{R}^N} \|x\|_1 \text{ such that } y = \Psi x$$

This basic formulation is formally defined as the basis pursuit (BP). The term was coined by Chen, Donoho, and Saunders (1998) because it was revealed that ℓ_1 norm has a tendency to locate the sparse solutions if ever they exist. Thanks to the strong linear programming theory and convexity, the set of well-developed, efficient, and powerful numerical solvers are available (Nocedal and Wright 1999). The results of this minimization with equality constraints are shown in Figure 8.10. It can be seen from Figure 8.10 (that 50-sparse signal $x \in \{0, 1, -1\}$ with 1024 samples) can be perfectly recovered with 220 number of measurements. But, it cannot be reconstructed properly if the numbers of measurements are inadequate (e.g., $M = 100$) (Figure 8.10b). Mean square error (MSE) is used as a measure to check the fidelity of the reconstructed sparse image. It is computed as follows for a one-dimensional (1D) signal.

$$MSE = \frac{1}{N} \left(\sum_{i=1}^{N} x_i - \hat{x}_i \right)^2 \tag{8.16}$$

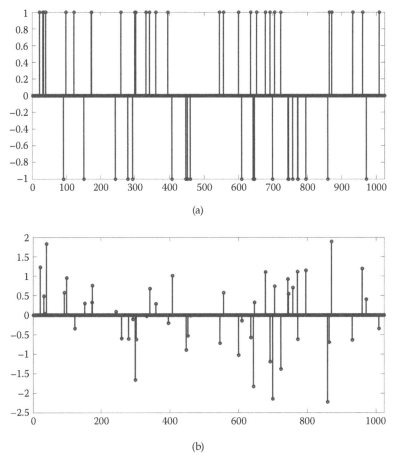

(a)

(b)

FIGURE 8.10
Results of ℓ_1 minimization with equality constraints (basis pursuit): (a) original one-dimensional signal: $N = 1024$, $k = 50$; (b) improperly reconstructed signal, $M = 100$, MSE = 0.0507, reconstruction time: 0.33 seconds; (c) properly reconstructed signal $M = 220$, MSE = 2.25×10^{-12}, reconstruction time: 0.27 seconds. (*continued*)

For the two-dimensional (2D) case, the definition of MSE is modified as follows:

$$MSE = \frac{1}{N \times N}\left(\sum_{i=1}^{N}\sum_{j=1}^{N} I(i,j) - \hat{I}(i,j)\right)^2 \tag{8.17}$$

where $x; I(i,j)$ & $\hat{x}; \hat{I}(i,j)$ represents the original and reconstructed sparse signal in 1D and 2D, $N; N \times N$ is the signal dimensionality for 1D and 2D cases, respectively. The results also indicate the total signal reconstruction

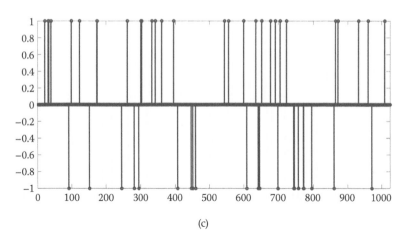

(c)

FIGURE 8.10 (*continued*)

Results of ℓ_1 minimization with equality constraints (basis pursuit): (a) original one-dimensional signal: $N = 1024$, $k = 50$; (b) improperly reconstructed signal, $M = 100$, MSE = 0.0507, reconstruction time: 0.33 seconds; (c) properly reconstructed signal $M = 220$, MSE = 2.25×10^{-12}, reconstruction time: 0.27 seconds.

time taken when measurements and measurement matrix are available. It is listed in seconds with each result.

The 2D checkerboard pattern reconstruction with basis pursuit equality constraints formulation is shown in Figure 8.11. The data normalization has been done on the 2D signal before it is used for the processing. The 2D signal will have a sparse gradient (Rudin, Osher, and Fatemi 1992). First-order approximation of the gradient in a 2D image I of size $N \times N$ is defined as follows:

$$D_h = I(i+1, j) - I(i, j) \tag{8.18}$$

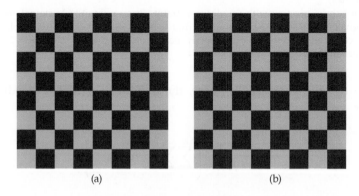

(a)   (b)

FIGURE 8.11

Two-dimensional signal reconstruction using ℓ_1 minimization with equality constraints (basis pursuit): (a) original image $N = 6400$ (80×80); (b) reconstructed image with $M = 2400$, MSE: 0.0 reconstruction time: 18.64 seconds.

where D_h is the first-order difference (gradient) in the horizontal direction, $I(i, j)$ represents a pixel at ith row and jth column:

$$D_v = I(i, j+1) - I(i, j), \tag{8.19}$$

where D_v is the first-order difference (gradient) in the vertical direction. Combining discrete gradient for a 2D image to get $D = \begin{bmatrix} D_h \\ D_v \end{bmatrix}$ and finding magnitude of D at each pixel in an image gives a more cumulative concrete measure (i.e., $\sum_{i,j} \|D\|_2$). Based on combined gradients, linear programs can be formulated for the sparse recovery.

The formulation can be slightly modified to incorporate the presence of noise ($y = \Psi x + n$) in the measurements. The new optimization problem is casted as

$$\hat{x} = \arg \min_{x \in \mathbb{R}^N} \|x\|_1 \text{ such that } \|y - \Psi x\|_2 \le \epsilon \tag{8.20}$$

where $\epsilon \ge \|n\|_2$ is the appropriate bound based on the noise magnitude and is a user-defined parameter. Then the solution $\hat{x}$ will be close to x, such that $\|\hat{x} - x\|_2 \le \gamma\epsilon$, where γ is a constant. These formulations are also known as the basis pursuit with the inequality constraint (BPIC). This type of optimization can be cased as a second order cone program, which is equivalent to the quadratically constrained program. The results of optimization for a 1D signal with inequality constraints are shown in Figure 8.12 with $\epsilon = 3 \times 10^{-3}$. It can be seen from Figure 8.12a–d that under the low noise condition, sparse signal can be faithfully reconstructed. With the increase in noise intensity, the signal recovery goes completely haywire. The results are depicted in Figure 8.12e–h on a second independent set of signals and measurements. The 2D signal reconstruction is shown in Figure 8.13 with $\epsilon = 3 \times 10^{-3}$. The signal is faithfully reconstructed with adequate numbers of measurement (Figure 8.13b) while with lower number of measurements, it is reconstructed improperly (Figure 8.13c).

One of the most popular techniques in the BPIC domain is to use the *Dantzig selector* (Candes and Tao 2005).

$$\hat{x} = \arg \min_{x \in \mathbb{R}^N} \|x\|_1 \text{ such that } \|\Psi^T(y - \Psi x)\|_\infty \le \zeta \tag{8.21}$$

In the above formulation ζ is a user-specified parameter. With respect to Gaussian noise $\mathcal{N}(0, \sigma^2)$, it is reformulated as follows:

$$\hat{x} = \arg \min_{x \in \mathbb{R}^N} \|x\|_1 \text{ such that } \|\Psi^T(y - \Psi x)\|_\infty \le \zeta\sqrt{\log N}\sigma \tag{8.22}$$

where, $\|\cdot\|_\infty$ is ℓ_∞ norm, which gives the largest member in a vector, and ζ controls the recovery. It relaxes the basis pursuit equality constraint. Then

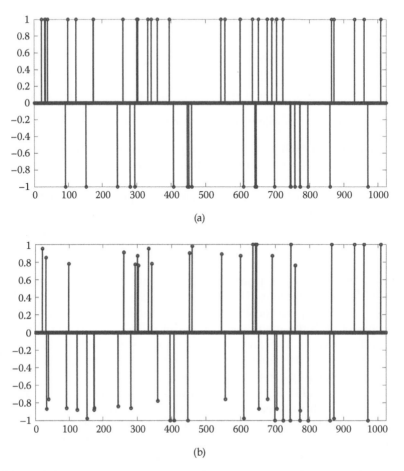

FIGURE 8.12
One-dimensional signal reconstruction using basis pursuit with inequality constraints: (a) one-dimensional signal x_1: $N = 1024$, $k = 50$; (b) reconstructed signal x_1 with $M = 220$ MSE = 2.02×10^{-4}, reconstruction time: 1.67 seconds; (c) noiseless measurements y_1; (d) noisy measurements y_1 with $\sigma = 0.005$; (e) another signal x_2: $N = 1024$, $k = 50$; (f) reconstructed signal x_2 with $M = 220$, MSE = 0.0488, reconstruction time: 1.85 seconds; (g) noiseless measurements y_2; (h) noisy measurements y_2 with $\sigma = 0.5$. (*continued*)

$\Psi^T(y - \Psi x)$ represents the correlation and ensures that $(y - \Psi x)$ is not heavily correlated with the columns of Ψ^T. These formulations are convex with the conic constraints; therefore, it can be solved using the linear programming. It also represents a quadratic program with polynomial complexity solver (Boyd and Vandenberghe 2004).

The majority of the CS solver attempts to solve the constrained versions of the formulation; however, it is interesting to look at the equivalent

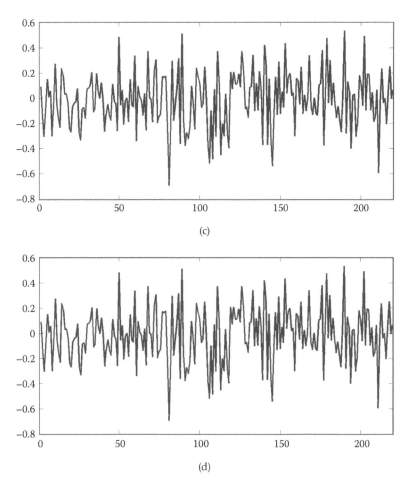

FIGURE 8.12 (*continued*)
One-dimensional signal reconstruction using basis pursuit with inequality constraints: (a) one-dimensional signal x_1: $N = 1024$, $k = 50$; (b) reconstructed signal x_1 with $M = 220$ MSE = 2.02×10^{-4}, reconstruction time: 1.67 seconds; (c) noiseless measurements y_1; (d) noisy measurements y_1 with $\sigma = 0.005$; (e) another signal x_2: $N = 1024$, $k = 50$; (f) reconstructed signal x_2 with $M = 220$, MSE = 0.0488, reconstruction time: 1.85 seconds; (g) noiseless measurements y_2; (h) noisy measurements y_2 with $\sigma = 0.5$. (*continued*)

unconstrained version, which has the following form (Boyd and Vandenberghe 2004; Beck and Teboulle 2009):

$$\hat{x} = \arg \min_{x \in \mathbb{R}^N} \|x\|_1 + \lambda \|y - \Psi x\|_2 \tag{8.23}$$

This formulation of the problem is known as basis pursuit denoising (BPDN). It is a Lagrangian relaxation of the quadratic solver. Various options of λ will

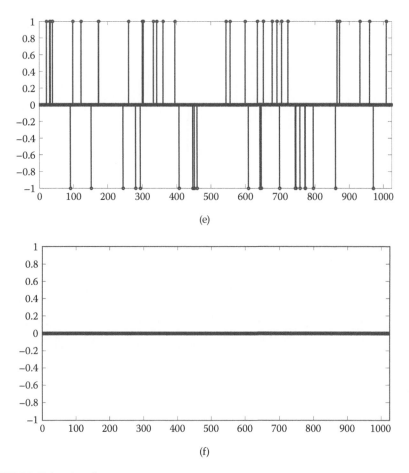

(e)

(f)

FIGURE 8.12 (*continued*)
One-dimensional signal reconstruction using basis pursuit with inequality constraints: (a) one-dimensional signal x_1: $N = 1024$, $k = 50$; (b) reconstructed signal x_1 with $M = 220$ MSE = 2.02×10^{-4}, reconstruction time: 1.67 seconds; (c) noiseless measurements y_1; (d) noisy measurements y_1 with $\sigma = 0.005$; (e) another signal x_2: $N = 1024$, $k = 50$; (f) reconstructed signal x_2 with $M = 220$, MSE = 0.0488, reconstruction time: 1.85 seconds; (g) noiseless measurements y_2; (h) noisy measurements y_2 with $\sigma = 0.5$. (*continued*)

yield the same form as the constrained formulation. The value of Lagrangian in general is unknown *a priori*. The excellent solvers for this problem are available and discussed by Becker, Bobin, and Candes (2011). However, one must note that ϵ is determined explicitly by the noise or quantization, making the constrained solver a more popular choice. Many constrained solvers are optimistic in a sense that they assume noise to be random and not bounded. One of the most popular choices in this is AWGN (additive white Gaussian noise) with the mean $\mu = 0$ and variance $\sigma^2 = 1$. The Bayesian

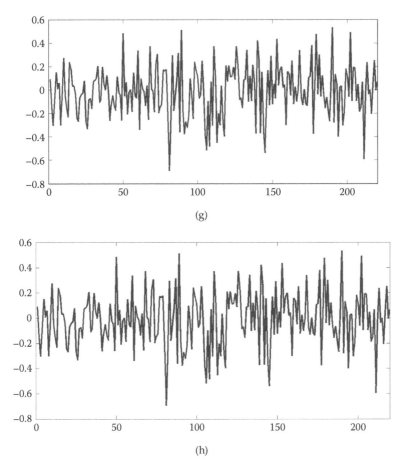

(g)

(h)

FIGURE 8.12 (*continued*)
One-dimensional signal reconstruction using basis pursuit with inequality constraints: (a) one-dimensional signal x_1: $N = 1024$, $k = 50$; (b) reconstructed signal x_1 with $M = 220$ MSE = 2.02×10^{-4}, reconstruction time: 1.67 seconds; (c) noiseless measurements y_1; (d) noisy measurements y_1 with $\sigma = 0.005$; (e) another signal x_2: $N = 1024$, $k = 50$; (f) reconstructed signal x_2 with $M = 220$, MSE = 0.0488, reconstruction time: 1.85 seconds; (g) noiseless measurements y_2; (h) noisy measurements y_2 with $\sigma = 0.5$.

estimation (Ji, Xue, and Carin 2008) poses a probability base prior, and then it is used together with the noise probability distribution. The results of signal reconstruction using the Dantzig selector for 1D and 2D signals are shown in Figures 8.14 and 8.15, respectively.

8.9.2 Greedy Sparse Recovery Algorithms

The greedy algorithms are based on the philosophy that during each step of iteration, a firm decision is made based on the locally optimal parameters.

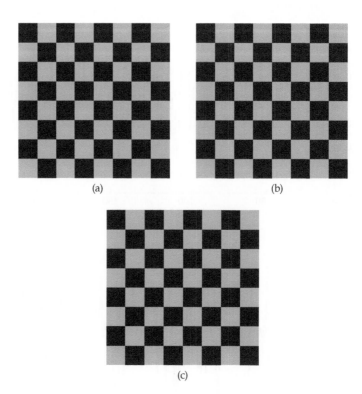

FIGURE 8.13
Two-dimensional signal reconstruction using basis pursuit with inequality constraints: (a) original image $N = 6400$ (80×80); (b) reconstructed image $M = 1600$, MSE: 0.25, reconstruction time: 16.27 seconds; (c) reconstructed image $M = 800$, MSE: 0.31, reconstruction time: 31.19 seconds.

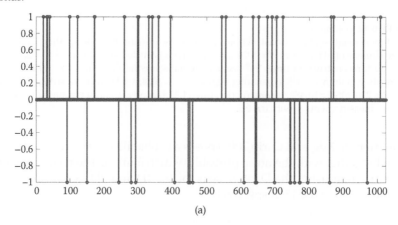

FIGURE 8.14
One-dimensional signal reconstruction using basis pursuit with inequality constraints and Dantzig selector: (a) original one-dimensional signal: $N = 1024$, $k = 50$; (b) reconstructed signal $M = 220$, MSE $= 1.74 \times 10^{-4}$ reconstruction time: 1.48 seconds, $\zeta = 3.00 \times 10^{-3}$. *(continued)*

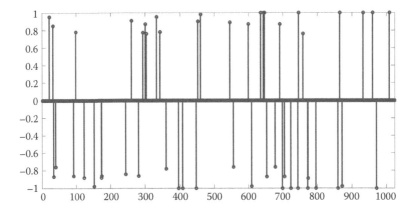

FIGURE 8.14 (*continued*)
One-dimensional signal reconstruction using basis pursuit with inequality constraints and Dantzig selector: (a) original one-dimensional signal: $N = 1024$, $k = 50$; (b) reconstructed signal $M = 220$, MSE $= 1.74 \times 10^{-4}$ reconstruction time: 1.48 seconds, $\zeta = 3.00 \times 10^{-3}$.

The first type of algorithm known as *pursuit* is a set of steps that iterates to construct an estimate of the signal. They initialize the estimate with a zero vector, find the set of nonzero components, optimize them, and add them to the estimate during each iteration. The pursuits approach provides very fast convergence; therefore, they are suitable for application requiring manipulation of the large dataset. Theoretical guarantee of apt recovery is weak in comparison with the other framework. *Matching pursuit* is one of the most straightforward pursuit algorithms (Mallat and Zhang 1993). It is summarized as shown in Algorithm 8.1 for ready reference.

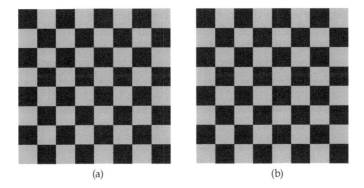

(a)   (b)

FIGURE 8.15
Two-dimensional signal reconstruction using basis pursuit with inequality constraints and Dantzig selector: (a) original image $N = 6400$ (80×80); (b) reconstructed image $M = 2400$, MSE: 0.25, reconstruction time 31.63 seconds, $\zeta = 5.00 \times 10^{-5}$.

Algorithm 8.1: Matching Pursuit
   Input: measurement y, measurement matrix Ψ, sparsity k
   Output: estimated signal $\hat{x}$, residual error r
   Initialize: $r_0 = y, \hat{x}_0 = 0$
   for $i = 1; i = i + 1;$ continue until the convergence condition is fulfilled

$$g_i = \Psi^T r_{i-1}$$

$$j_i = \arg\max_j \frac{|g|_j^i}{\|\Psi_j\|_2}$$

$$\hat{x}_j^i = \hat{x}_j^{i-1} + \frac{g_j^i}{\|\Psi_j\|_2^2}$$

$$r_i = r_{i-1} - \frac{\Psi_j g_j^i}{\|\Psi_j\|_2^2}$$

end for

$$return \;\; \hat{x} \leftarrow \hat{x}_i \qquad\qquad\blacksquare$$

Note that Ψ_j represents the jth column of matrix Ψ. The approximation is augmented by selecting each column of measurement matrix at a time, and only the coefficient associated with the selected column is updated. The updates $\hat{x}_j^i = \hat{x}_j^{i-1} + \frac{g_j^i}{\|\Psi_j\|_2^2}$ minimize the approximation cost $\|y - \Psi\hat{x}_i\|_2^2$ for the selected coefficients. It may happen that MP may go over the same columns of a measurement matrix for refinement of the solution. In MP the minimization is performed on the most recent coefficients of the matrix elements. One can also define a hard stopping criteria by fixing the strength of the residual r. MP computations are dominated by the step $g_i = \Psi^T r_{i-1}$. Therefore, to speed up the computations, many times Ψ is based on Fourier transform, which can be implemented using fast Fourier transform (FFT). Figure 8.16 showcases the typical 1D signal reconstruction using matching pursuit for the noiseless and noisy scenarios. It can be observed that signal recovery falters as the noise level is increased but gives nearly perfect recovery in the noiseless scenario.

1. Orthogonal matching pursuit

   A more sophisticated implementation of MP is in the form of orthogonal matching pursuit (OMP) (Mallat, Davis, and Zhang 1994). The algorithm is summarized as follows:

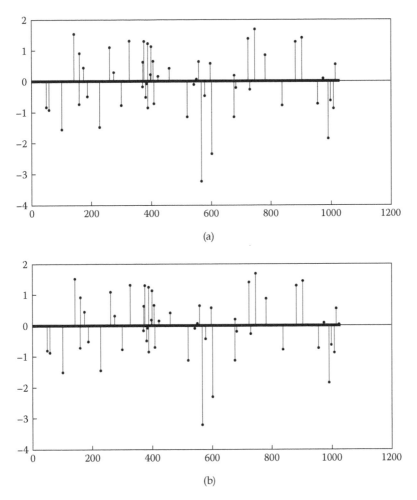

FIGURE 8.16
Matching pursuit one-dimensional signal reconstruction: (a) original signal; (b) noiseless one-dimensional reconstruction; (c) medium noisy one-dimensional reconstruction; (d) heavily noisy one-dimensional reconstruction. (*continued*)

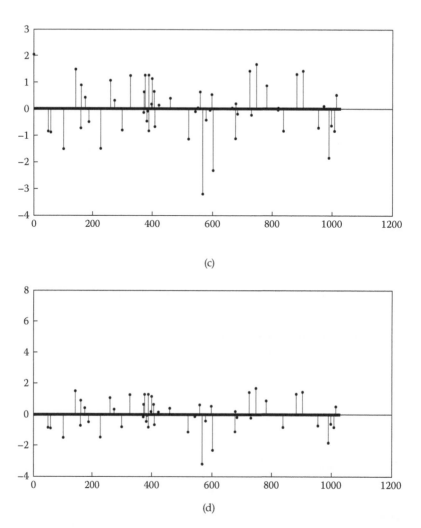

(c)

(d)

FIGURE 8.16 (*continued*)
Matching pursuit one-dimensional signal reconstruction: (a) original signal; (b) noiseless one-dimensional reconstruction; (c) medium noisy one-dimensional reconstruction; (d) heavily noisy one-dimensional reconstruction.

Algorithm 8.2: Orthogonal Matching Pursuit (OMP)
  Input: measurement y, measurement matrix Ψ, sparsity k
  Output: estimated signal $\hat{x}$, residual error r
  Initialize: $r_0 = y, \hat{x}_0 = 0$
  for $i = 1; i = i + 1;$ continue until convergence condition is fulfilled

$$g_i = \Psi^T r_{i-1}$$

$$j_i = \arg\max_j \frac{|g|_j^i}{\|\Psi_j\|_2}$$

$$Supp_i = Supp_{i-1} \cup j_i$$

$$\hat{x}^i_{Supp} = \Psi^\dagger_{Supp} y$$

$$r_i = y - \Psi\hat{x}_i$$

end for

$$return \ \hat{x} \leftarrow \hat{x}_i \qquad \blacksquare$$

In OMP $\hat{x}$ is restructured by projecting measurement y orthogonally onto the columns of Ψ, which is associated with the support $Supp_i$. Then $\dagger$ indicates the Moore-Penrose inverse operator. In OMP, all the coefficients with support are involved in the minimization:

$$\hat{x}^i_{Supp} = \arg\min_{x^i_{Supp}} \|y - \Psi_{Supp}\tilde{x}_{Supp}\|_2^2 \qquad (8.24)$$

In OMP, the minimization is done on the selected coefficients at iteration i, and it never goes over the same coefficients again, unlike MP. Residual r is always orthogonal to selected coefficients. OMP is more computationally intensive than MP, and it gives superior recovery guarantees. Computational cost and storage requirement of a single iteration of OMP is very high for a big dataset. It will be very slow due to selection of k atoms of Ψ to estimate measurement y. Various forms of factorization can be done to solve and improve the speed of the least square problems. Figure 8.17 indicates 1D signal recovery by the orthogonal matching pursuit algorithm. It reconstructs the signal flawlessly with practically MSE = 0. The signal can be uniquely represented by 256 measurements, which is only one-quarter of the signal size. The signal reconstruction time is also pretty good. However, the robustness of the recovery deteriorates with the increase in noise level (Figure 8.17c,d).

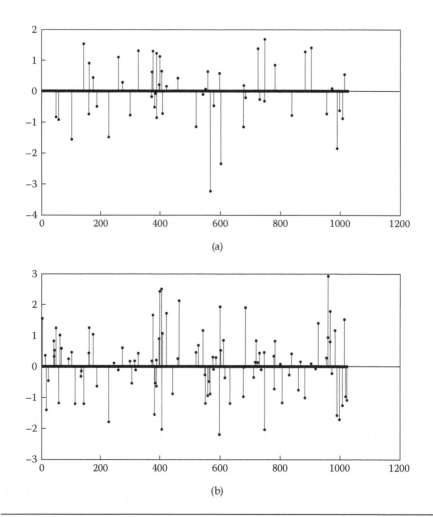

(a)

(b)

(8.17a) Original 1D signal $N = 1024$, $k = 51$	(8.17b) Noiseless 1D signal reconstruction, $M = 256$, MSE $= 5.83 \times 10^{-16}$, reconstruction time = 0.043 seconds
(8.17c) Medium noisy 1D signal reconstruction, $M = 256$, MSE $= 3.28 \times 10^{-1}$, reconstruction time = 0.033 seconds, noise $\sigma = 0.12$	(8.17d) Heavily noisy 1D signal reconstruction, $M = 256$, MSE $= 9.04 \times 10^{-1}$, reconstruction time = 0.012 seconds, noise $\sigma = 0.38$

FIGURE 8.17
OMP one-dimensional signal reconstruction: (a) original signal; (b) noiseless one-dimensional reconstruction; (c) medium noisy one-dimensional reconstruction; (d) heavily noisy one-dimensional reconstruction. (*continued*)

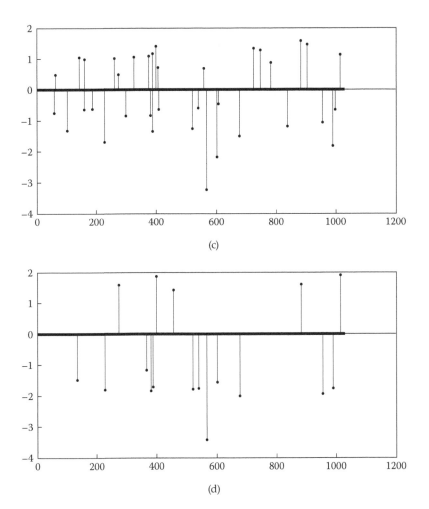

FIGURE 8.17 (*continued*)
OMP one-dimensional signal reconstruction: (a) original signal; (b) noiseless one-dimensional reconstruction; (c) medium noisy one-dimensional reconstruction; (d) heavily noisy one-dimensional reconstruction.

1. Thresholding algorithms

The thresholding algorithms are sort of an optimized solution between the theoretically stronger ℓ_1 minimization and easier to implement greedy pursuit algorithms. Thresholding algorithms provide strong recovery guarantees with an efficient implementation. In this chapter, we will consider iterative hard thresholding (IHT) (Blumensath and Davies 2008; Portilla 2009) for discussions.

2. Iterative hard thresholding (IHT)

It is a greedy algorithm that solves the confined approximation in the CS sparse recovery. It is formulated as

$$\min_{\tilde{x}} \|y - \Psi\tilde{x}\|_2^2 \quad \text{subject to} \quad \|\tilde{x}\|_0 \leq k. \tag{8.25}$$

This problem is solved by introducing a substitute function as follows:

$$\mathcal{L}_k^S(\tilde{x}, z) = \mu \|y - \Psi\tilde{x}\|_2^2 - \mu \|\Psi\tilde{x} - \Psi z\|_2^2 + \|\tilde{x} - z\|_2^2 \tag{8.26}$$

The minimization of this equation under the constraint $\|\tilde{x}\|_0 \leq k$ is obtained as

$$\hat{x} = \mathfrak{H}_k(z + \mu\Psi^T(y - \Psi z)) \tag{8.27}$$

where $\mathfrak{H}_k$ is the nonlinear projection that preserves the largest k nonzero coefficients and sets all other coefficients to zero, and μ is a step size. By setting $z = \hat{x}_i$ it gets converted into the IHT algorithm which is described as follows:

Algorithm 8.3: Iterative Hard Thresholding (IHT)
Input: measurement y, measurement matrix Ψ, sparsity k, step size μ
Initialize: $\hat{x}_0 = 0$
for $i = 1; i = i + 1$; continue until convergence condition is fulfilled

$$\hat{x}_{i+1} = \mathfrak{H}_k(\hat{x}_i + \mu\Psi^T(y - \Psi\hat{x}_i))$$

end for   ■

It is computationally efficient and easier to implement. The storage requirements are small and multiplication of matrix and vectors can be done efficiently. Theoretically, the method guarantees recovery, matches in comparison with the other convex optimization techniques, and can be adapted to incorporate signal structure beyond sparsity. This interesting feature makes it suitable for non-sparse-signal recovery as well. The thresholding step requires sorting. Figure 8.18 indicates the 1D signal construction for the IHT-based sparse recovery algorithm. For noisy cases, the recovery is better as compared to OMP. However, signal reconstruction time increases as the noise intensity is increased, and the signal fidelity degrades significantly with increase in noise variance.

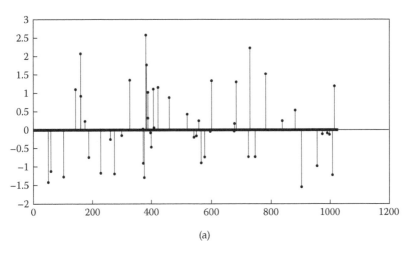

(a)

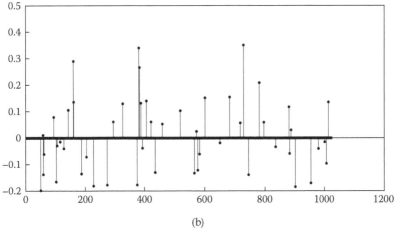

(b)

(8.18a) Original 1D signal $N = 1024, k = 50$	(8.18b) Noiseless 1D signal reconstruction, $M = 256$, MSE $= 3.99 \times 10^{-2}$, reconstruction time $= 0.060$ seconds
(8.18c) Medium noisy 1D signal reconstruction, $M = 256$, MSE $= 4.28 \times 10^{-2}$, reconstruction time $= 3.66$ seconds, noise $\sigma = 0.3$	(8.18d) Heavily noisy 1D signal reconstruction, $M = 256$, MSE $= 4.57 \times 10^{-2}$, reconstruction time $= 3.77$ seconds, noise $\sigma = 0.9$

FIGURE 8.18
IHT one-dimensional signal reconstruction: (a) original signal; (b) noiseless one-dimensional reconstruction; (c) medium noisy one-dimensional reconstruction; (d) heavily noisy one-dimensional reconstruction. (*continued*)

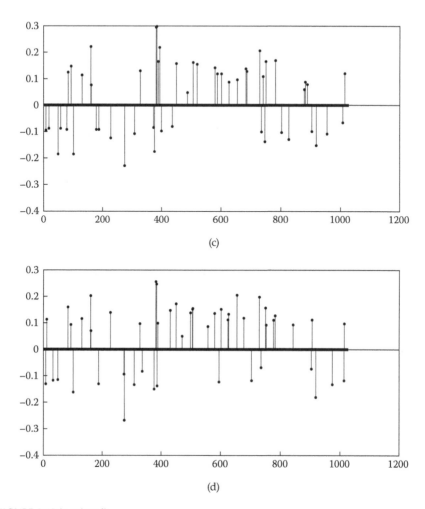

FIGURE 8.18 (*continued*)

IHT one-dimensional signal reconstruction: (a) original signal; (b) noiseless one-dimensional reconstruction; (c) medium noisy one-dimensional reconstruction; (d) heavily noisy one-dimensional reconstruction.

8.10 Recovery Guarantees for Greedy Algorithms

Throughout this chapter we discussed various cornerstone results provided by the researchers on recovery guarantees of the sparse recovery ℓ_1 minimization algorithms. Now we will carry forward with respect to greedy algorithms. While convergence is clearly very important criteria, in CS it is more important to recover a sparse or nearly sparse signal. The weakness of these algorithms lies in their theoretical recovery guarantee. It has been shown (Donoho

2006) that for certain random matrixes, OMP fails to provide uniform recovery guarantee for all k-sparse vectors $x \in \mathbb{C}_k$. This is due to the *enforced* regime on OMP that it must select *only* correct elements from the measurement matrix. The uniform recovery guarantee by sparse pursuit algorithm (OMP) in terms of RIP has been given by Davenport and Wakin (2009).

Theorem 8.5 (Davenport and Wakin 2009): Let $y = \Psi x + e$ be the measurements in error and algorithm stops after selecting k nonzero elements, then x_k is the best k term approximation of x if the following is satisfied if there exists a constant C such that

$$\|\hat{x}_i - x\|_2 \leq C \left(\|(x - x_k)\|_2 + \frac{\|(x - x_k)\|_1}{\sqrt{k}} + \|e\|_2 \right) \qquad (8.28)$$

where $\hat{x}_i$ is the estimate during the last iteration i. This bound on the error is optimal up to a fixed value, and it is similar to ℓ_1 minimization. The theorem proposed by Davenport and Wakin (2009) is very conservative bound and may not indicate the general achievable bounds by greedy algorithms. It has been shown by Tropp and Gilbert (2006) that sparse signal can be successfully recovered using OMP for $M = \mathcal{O}(k\log(n))$ measurements. Here Ψ is drawn at random for a suitable random set. ∎

Theorem 8.6 (Tropp and Gilbert 2006): Suppose that signal $x \in \mathbb{C}_k$ *and* $\mathbb{R}^N$. $\Psi \in \mathbb{R}^{M \times N}$ is drawn randomly from the i.i.d. Gaussian or Bernoulli distribution. Given measurement $y = \Psi x$, choosing $m \geq Ck\log(\frac{n}{\sqrt{\delta}})$, where C is a constant based on Ψ, then OMP can reconstruct the signal with probability $1 - \delta$. ∎

This theorem showcases that by choosing Ψ independently of the signal, OMP will exhibit good performance recovery even under the case when M scales with k.

In the case of IHT, convergence criteria are based on the step size μ. It is dependent on measurement matrix Ψ property, namely, ρ_{2k}, defined as the minimum quantity such that

$$\|\Psi(x_1 - x_2)\|_2^2 \leq \rho_{2k} \|(x_1 - x_2)\|_2^2 \qquad (8.29)$$

holds for all k-sparse vectors x_1 and x_2. It is clear that $\rho_{2k} \leq (1 + \delta_{2k}) \leq \|\Psi\|_2^2$, where δ_{2k} is an RIP constant. The convergence results derived by Blumensath and Davies (2008) are stated as follows.

Theorem 8.7 (Blumensath and Davies 2008): If $\rho_{2k} < \frac{1}{\mu}$, $k \leq M$, and assume Ψ to be full-rank matrix, then the sequence $\{\hat{x}_i\}$, defined by the IHT converges to a local minimum of $\min_{\tilde{x}} \|y - \Psi\tilde{x}\|_2^2$ subject to $\|\tilde{x}\|_0 \leq k$. ∎

One of the fundamental properties of IHT is that the sparse recovery guarantee holds wherever RIP holds true. This performance is similar to the convex optimization.

We will now look at developing an application in the domain of watermarking, which will utilize the principles of compressive sensing.

8.11 Semifragile Watermarking Technique Based on Compressed Sensing

A manipulated image, video, or a piece of audio can wreck havoc with an individual or society at large. Therefore, it is mandatory to protect the integrity of these digital signals. There are several applications wherein it is important to protect the digital image integrity (e.g., in medical imaging where a loss of integrity may result in a faulty diagnosis or an intentional manipulation of a photographic evidence in a criminal trial suit may lead to a wrong judgment and conviction). The semifragile technique verifies the integrity of the image because watermark resides within the image and undergoes the same manipulations as the image. By comparing the extracted watermark against the reference watermark in terms of a distance metric, one may detect the loss of integrity. However, there are certain manipulations that are nonmalicious while few are malicious. Therefore, semifragile watermarking schemes are designed to differentiate between malicious and nonmalicious manipulations. They flag off loss of integrity only in the case of malicious attacks and thus justify the name *semifragile*.

The proposed application scheme in this chapter uses a watermark that is resistive to nonmalicious attacks of JPEG compression, downsampling, and fragile to malicious attacks like deliberate insertion of data into the image. In recent years, compressive sensing–based image authentication mechanisms have been gaining popularity. CS allows reconstruction of a sparse signal from its samples, which are sampled at a rate much lower than the Nyquist rate. By using CS theory the detection of tinkering to the multimedia content can be obtained using the following techniques: (1) image hashing–based techniques or (2) semifragile or content-fragile watermarking. An image hashing method requires succinct feature representation (hash) of digital content and has the following requirements (Kang, Lu, and Hsu 2009): (1) perceptual transparency, (2) conflict resistivity, (3) one-way directionality, and (4) randomness. Kang, Lu, and Hsu (2009) propose an image hashing technique for the watermark . Here, the given image is first downsampled and random projection onto a lower dimension measurement vector is obtained. Each component in the measurement vector is then quantized to generate the hash vector, which is followed by entropy encoding. For authentication, a similar hash-generation procedure is applied using the same secret key. Lin, Varodayan, and Girod (2009) propose a distributed source coding (DSC)–based hashing scheme. DSC-encoded random projections of image coefficients are quantized to compute the hash values that are used as watermark. A DSC-based decoder is used to check the integrity. Similar schemes have also been proposed by Tagliasacchi et al. (2008, 2009).

Another set of techniques used for tampering detection is based on the semifragile watermark. Unlike the image hashing technique, the semifragile method embeds the watermark directly into the digital content in order to ensure the integrity of digital content. Zhang et al. (2011) proposed a tampering detection, localization, and recovery of original coefficients assuming that the tampering is sparse. The embedded watermark is image dependent and the method exploits the sparseness in the discrete cosine transform (DCT) domain. The extracted watermark from the nontampered area is used for localization and the content recovery is using the CS-based approach. After image decimation, Tagliasacchi et al. (2009) uniformly quantized random projections. A low-density parity check code is applied on the quantized values to form the signature that is then embedded as a robust watermark. Recovery of the watermark leads to a recovery of projections, which in turn are used to obtain the estimate of the distortion in the image. Sheikh and Baraniuk (2007) present a CS-based watermark detection in a DCT domain . The method relies on the fact that for natural images, coefficients in the transform domain are sparse. An additive watermarking scheme is used for insertion of the watermark in the middle frequency coefficients. The watermark is recovered using the CS-based scheme under the condition of sparsity. The robustness of the scheme against Poisson noise attack has been demonstrated by the authors.

We develop a semifragile technique in which the watermark is robust to operations such as JPEG compression and downsampling, but fragile to malicious operations like deliberate addition of data to image. Sheikh and Baraniuk (2007) proposed a CS-based method for watermarking, wherein no distinction has been made between such attacks. We insert the watermark in the mid-frequency coefficients using block-wise DCT that ensures better perceptual fidelity. Since it performs local embedding of watermark, an additional sparsification in DCT domain is achieved by using the JPEG quantization matrix of desired quality factor. The schematic of the proposed approach is shown in Figures 8.19 and 8.20.

We propose a semifragile watermarking scheme considering a noisy channel scenario as presented by Tropp and Wright (2010), making it an overdetermined set of equations. Consider that a sparse error vector e corrupts a vector x as follows:

$$y = \Psi x + e \tag{8.30}$$

where $\Psi \in \mathbb{R}^{M \times N}$ with $M > N$. The x is recovered as follows:

1. Construct matrix $\mathcal{F}$ such that $\mathcal{F}\Psi = 0$
2. Premultiply the corrupted y by $\mathcal{F}$; that is,

$$\hat{y} = \mathcal{F}y = \mathcal{F}(\Psi x + e) = \mathcal{F}e \tag{8.31}$$

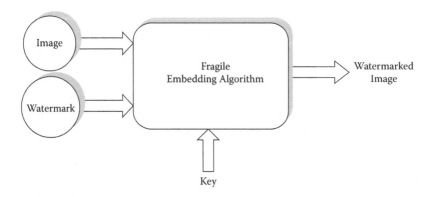

FIGURE 8.19
Watermarking encoder.

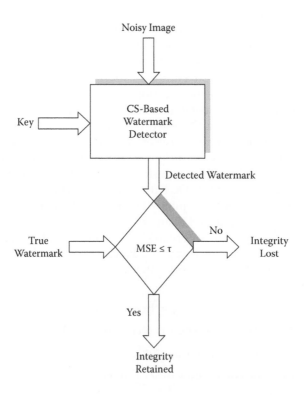

FIGURE 8.20
Watermarking detector.

3. Estimate $\hat{e}$ using the CS formulation:

$$\min_{e} |e|_1 \quad \text{subject to} \quad \hat{y} = \mathcal{F}e \qquad (8.32)$$

4. Finally, recover $\hat{x}$—that is,

$$\hat{x} = \Psi^{\dagger}(y - \hat{e}) \qquad (8.33)$$

where † represents the Moore-Penrose inverse. Since Ψ is full-column rank, its right pseudo inverse exists and is given as $\Psi^{\dagger} = \Psi^T(\Psi\Psi^T)^{-1}$.

Equations (8.31) to (8.33) summarize our CS-based watermarking scheme. The given watermark (x) is encoded as a sequence (refer to Equation 8.30) by multiplying it with the Gaussian measurement matrix (Ψ) and the resulting vector is added to sparse transform domain coefficients (e). The watermark can be recovered by using Equation (8.33) after determining the sparse coefficients as in Step 3.

In the proposed algorithm, we consider the transform domain for watermark insertion because it is more robust to attacks, adaptive to watermarking operations, and also useful for obtaining the sparse representation of an image.

8.11.1 Watermark Embedding

Our method for watermark insertion is as follows. Given a confidential watermark vector w drawn from as i.i.d. Gaussian distribution, we first encode it as $r = \Psi w$, where $\Psi \in \mathbb{R}^{M \times N}$ is a Gaussian matrix drawn from a seed $\mathfrak{K}_e$ known only to the embedder and the decoder. This ensures security as the decoder requires the same seed to generate Ψ. The generated vector r represents the uncorrupted measurement vector in the CS term and also can be considered as a code word in a channel coding sense. To form the watermarked image, the image is divided into 8×8 nonoverlapping blocks, and the DCT is computed on each block. An 8×8 JPEG quantization matrix of desired quality factor Q_{design} is then used on these blocks (divide each DCT coefficient by the corresponding entry in the quantization matrix) to obtain the quantized middle frequency DCT coefficients in which the insignificant coefficients are rounded to zero as done in the theory of compression. This step brings additional sparseness in the middle frequency coefficients. Now the middle frequency DCT coefficients e in each block are added to r in order to obtain the watermarked image in the transform domain (i.e., we obtain the following watermarked vector, $y = r + e = \Psi w + e$).

This model is in accordance with Equation (8.30), where w is the watermark vector and e is the sparse error vector. Here, y represents the noisy measurement vector. It should be noted that insertion of a watermark into low-frequency coefficients causes loss of perceptual fidelity while embedding watermark in high-frequency coefficients distorts the edge details. This motivates us to use the middle frequency coefficients e for embedding the watermark. They also achieve an optimal compromise between the robustness (due to nonmalicious attack) and fragility (against malicious attack). Finally, each 8×8 block is multiplied by Q_{design}, which will restore low- and high-frequency coefficients, and the inverse DCT is taken to obtain the watermarked image. We mention here that although the image is manipulated in the DCT domain, the sparseness in e causes negligible loss in the image. Since the sparsification in the transform domain is carried out in a way similar to that used in JPEG image compression, our scheme can be used for inserting watermark directly into the JPEG compressed image. Moreover, this semifragile scheme can be extended to any transform domain yielding sparse vector representation.

8.11.2 Watermark Detection

In order to check integrity by detecting the watermark, we first downsample and transform the image into the DCT domain. Once again we use 8×8 nonoverlapping blocks as was done while embedding the watermark. Dividing the middle frequency DCT coefficients by Q_{design} gives us the vector y^* using the same seed $\mathfrak{K}_e$ to generate the $\Psi \in \mathbb{R}^{M \times N}$ matrix; the sparse vector e can be estimated by referring to Equation (8.32). Finally, the corrupted watermark is estimated as

$$\hat{w} = \Psi^\dagger (y^* - \hat{e}) \tag{8.34}$$

8.11.3 Results and Observations

We now show experiments on the proposed watermarking scheme. The integrity of images is verified by detecting the presence of a watermark in it. The watermark vector has a length of $N = 100$. Based on the CS theory, watermark measurement vector of length $M = 1280$ is generated. The image of size 256×256 is divided into 8×8 nonoverlapping 64 blocks. After taking DCT on each block we selected 20 middle frequency coefficients per block resulting in a total of $64 \times 20 = 1280$ coefficients, then the block-wise watermark measurement vector is added to the DCT coefficients. The watermarked images are shown in Figure 8.21.

The perceptual fidelity of an image is quantified using the document-to-watermark (DWR) ratio and visual information fidelity (VIF) (Sheikh and Bovik 2006). Both measures are computed between the images with and without watermark. Table 8.2 indicates these measures for different images. It shows that DWR values are high and VIF values are close to one. It means that the perceptual fidelity of the image is very good.

For evaluating integrity, we downsample the image by a factor of four and extract the watermark. We compare the MSE between the original

(a)

(b)

FIGURE 8.21
Watermarked images: (a) leaves; (b) label; (c) kid; (d) Ganpati. (*continued*)

(c)

(d)

FIGURE 8.21 (*continued*)
Watermarked images: (a) leaves; (b) label; (c) kid; (d) Ganpati.

TABLE 8.2

Performance Comparison Using DWR and VIF for Watermarked
Images

Image	DWR (dB)	VIF
Leaves	45.6	0.976
Label	47.4	0.976
Kid	48.7	0.975
Ganpati	50.3	0.975

TABLE 8.3

MSE from Downsampled and JPEG Compressed Images

Image	MSE	
	Downsampled	**JPEG Compressed**
Leaves	9.85×10^{-5}	2.0×10^{-3}
Label	2.84×10^{-4}	9.93×10^{-4}
Kid	1.6×10^{-5}	9.4×10^{-3}
Ganpati	3.85×10^{-5}	5.31×10^{-4}

watermark w and the estimated watermark $\hat{w}$ to decide the integrity of an image. It preserves the integrity if the MSE $(w, \hat{w}) < \tau_{mse}$, where τ_{mse} represents a fixed threshold which is predetermined by posing it as a hypothesis testing problem. The estimated value of τ_{mse} in our experimentation is 0.1284. The watermark should be robust to nonmalicious attacks such as downsampling and compression. The experimentation of extracting watermark from the downsampled version simulates the downsampling attack on the watermarking system. The preservation of integrity in spite of downsampling indicates that the watermarking method is robust against downsampling.

In many of the applications, images have to be subjected to compression whenever we want to store them in a limited size memory or transmit them through band-limited channels. In such situations, we must be able to recover the embedded watermark if we want to preserve the integrity. In order to show the robustness of the semifragile technique, we test the scheme against JPEG compression. We use $Q_{design} = 50$ while inserting the watermark measurement in the image. The method recovers the watermark if the image is subjected to JPEG compression with quality factor $Q > 50$. We use the value of τ_{mse} as 0.1284 in our experimentation. Table 8.3 indicates the results of our experimentation. It can be seen that $MSE < \tau_{mse}$ for all the test images, flagging integrity preservation.

The watermark should be fragile to the malicious manipulations like substantial editing of an image. This editing manipulation is simulated by deliberately tampering 25% spatial locations of watermarked image. Under this editing, 25% grayscale values are replaced with a fixed value of 128. Table 8.4

TABLE 8.4

MSE after Editing Attack

Image	MSE
Leaves	0.2897
Label	0.8422
Kid	3.1326
Ganpati	0.2514

indicates the outcome of our experimentation. It shows that MSE $> \tau_{mse}$ for the test images, indicating integrity loss.

Thus, the semifragile watermarking technique based on CS can be successfully deployed for checking the image integrity and the application can be used for content authentication.

8.12 Summary

Compressive sensing is one of the most exciting domains of the modern era with deeply rooted mathematical theory derived from various disciplines of signal processing, statistics, probability theory, computer science, optimization, and linear programming. In this chapter, we have visited these basic theories behind the CS and noted that they are interwoven strongly. This chapter is aimed at both the novice and seasoned practitioner in the domain of CS. For the novice, this chapter provides a comprehensive overview of the basic theory, and it acts as a one-point reference for the seasoned researcher. This chapter has been deliberately placed at the end of the book, with emphatic coverage on the conventional compression covered earlier in the book. This is done with the purpose so that readers are motivated toward a new dimensionality of thinking. The author has attempted to cover the basics of CS from the theoretical framework perspective and also by using the algorithmic approach. The authors also weave an application of fragile domain watermarking using CS so that readers are motivated to put the theory into practice. The aim is to develop an application within the basic framework of CS covered earlier in the chapter.

The current focus in the CS community is to look beyond the random measurement matrix and assumed sparsity constraints in the signal model. The measurement matrices can be more structured, and they incorporate the underlying hardware dynamics. It has emerged from the way in which the signals are acquired for the real-world applications. Structured matrices design has demanded the attention due to the fact that the multiplication cost is very high when random matrices are multiplied with a signal of very high dimensionality. Many times the sensing matrices are deployed for the specific applications. It includes the device measurement ability and degree of measurement freedom by the sensing devices. CS all along has been driven by the desire to build a sampling mechanism that captures the signal below the Nyquist rate. On the other hand, actual hardware considerations require an elaborate signal model that helps in minimizing the sampling rate.

The other area of movement in the research domain of CS is developing better signal models that look ahead of sparsity and are applicable to a more broader class of the signal. The earlier work in the CS was developed considering the finite dimensional signals; however, now CS researchers are developing models for continuous time signal with infinite dimensions. These models can further help in pushing the number of measurement M as close as possible to the sparsity k; thus, higher compression can be obtained. One has to note limitations of the standard sparsity model in describing the infinite dimensional signal and its demands for development of new models. It has led to the development of low rate sampling methods like finite rate innovation (FRI) (Dragotti, Vetterli, and Blu 2007) and Xampling framework (Mishali et al. 2011). Structure in the models of analog signal and its sensing can help in sampling it at a sub-Nyquist rate, and that is the prime goal of researchers in the CS domain in the years to come.

Bibliography

A. Aldroubi, X. Chen, and A. M. Powell, Perturbations of measurement matrices and dictionaries in compressed sensing. *Appl. Comput. Harmon. Anal.*, 33(2): 282–291, 2012.

W. Alltop, Complex sequences with low periodic correlations. *IEEE Trans. Inform. Theory*, 26(3): 350–354, May 1980.

W. Bajwa, R. Calderbank, and S. Jafarpour, Why Gabor frames? Two fundamental measures of coherence and their role in model selection. *IEEE Jour. Commu. and Netw.*, 12(4): 289–307, August 2010.

R. Baraniuk, Compressive sensing. *IEEE Sig. Proc. Mag.*, 24(4): 118–124, 2007.

R. G. Baraniuk, M. Davenport, R. A. DeVore, and M. Wakin, A simple proof of the restricted isometry property for random matrices. *Constr. Approx.*, 28: 2008.

A. Beck and M. Teboulle, A fast iterative shrinkage-thresholding algorithm for linear inverse problems. *SIAM Jour. Imag. Sci.*, 2(1): 183–202, 2009.

S. Becker, J. Bobin, and E. J. Candès, NESTA: A fast and accurate first order method for sparse recovery. *SIAM Jour. Imag. Sci.*, 4(1): 1–39, 2011.

T. Blumensath and M. Davies, Iterative thresholding for sparse approximations. *J. Fourier Anal. Appl.*, 14(5): 629–654, 2008.

B. Bodmann, P. Casazza, and G. Kutyniok, A quantitative notion of redundancy for finite frames. *Appl. Comput. Harmo. Anal.*, 30(3): 348–362, 2011.

P. Boufounos, H. Rauhut, and G. Kutyniok, Sparse recovery from combined fusion frame measurements. *IEEE Trans. Inf. Theory*, 57(6): 3864–3876, 2011.

S. Boyd and L. Vandenberghe, *Convex Optimization*, Cambridge University Press, New York, 2004.

A. M. Bruckstein, D. L. Donoho, and M. Elad, From sparse solutions of systems of equations to sparse modeling of signals and images. *SIAM Rev.*, 51(1): 34–81, February 2009.

T. T. Cai, G. Xu, and J. Zhang, On recovery of sparse signal via ℓ_1 minimization. *IEEE Trans. Inf. Theory*, 55(7): 3388–3397, July 2009.

E. J. Candes, J. Romberg, and T. Tao, Robust uncertainty principles: Exact signal reconstruction from highly incomplete frequency information. *IEEE Trans. Inf. Theory*, 52(2): 489–509, December 2006.

E. J. Candes, J. Romberg, and T. Tao, Stable signal recovery from incomplete and inaccurate measurements. *Comm. Pure Appl. Math.*, 59: 1207–1223, 2006.

E. J. Candes and T. Tao, The Dantzig selector: Statistical estimation when p is much larger than n. *Ann. Statist.*, 35(6): 2313–2315, 2005.

E. J. Candes and T. Tao, Decoding by linear programming. *IEEE Trans. Inf. Theory*, 51(12): 4203–4215, 2005.

E. J. Candes and T. Tao, Near optimal signal recovery from random projections: Universal encoding strategies. *IEEE Trans. Inf. Theory*, 52(12): 5406–5425, December 2006.

E. J. Candes and T. Tao, The power of convex relaxation: Near-optimal matrix completion. *IEEE Trans. Inform. Theory*, 56(5): 2053–2080, 2009.

P. Casazza and G. Kutyniok, *Finite Frames*, Birkhauser, Boston, MA, 2012.

S. S. Chen, D. L. Donoho, and M. A. Saunders, Atomic decomposition by basis pursuit. *SIAM J. Scientific Computing*, 20(1): 33–61, 1998.

X. Chen, R. Wang, and H. Wang, A null space property approach to compressed sensing with frames. *Proc. of 10th International Conference on Sampling Theory and Applications*, July 1, 2013.

Y. Chi, L. L. Scharf, A. Pezeshki, and R. Calderbank, Sensitivity to basis mismatch in compressed sensing. *IEEE Trans. Signal Process*, 59(5): 2182–2195, May 2011.

A. Cohen, W. Dahmen, and R. DeVore, Compressed sensing and best k-term approximation. *J. Am. Math Soc.*, 22: 211–231, 2009.

M. Davenport, Random observations on random observations: Sparse signal acquisition and processing, PhD thesis, Rice University, 2010.

M. A. Davenport and M. B. Wakin, Analysis of orthogonal matching pursuit using the restricted isometry property. *IEEE Trans. Infor. Theor*, 56(9): 4395–4401, 2009.

D. L. Donoho, Compressed sensing. *IEEE Trans. Inf. Theory*, 52(4): 1289–1306, September 2006.

D. L. Donoho, For most large underdetermined systems of linear equations the minimal l-norm solution is also the sparsest solution. *Commun. Pure Appl. Math.*, 59(6): 797–829, 2006.

D. L. Donoho and M. Elad, Optimally sparse representation in general (nonorthogonal) dictionaries via L1 minimization. *Proc. Nat. Aca. Sci.*, 100: 2197–2202, March 2003.

D. L. Donoho and X. Huo, Uncertainty principles and ideal atomic decomposition. *IEEE Trans. Inf. Theory*, 47(7), 2845–2862, November 2001.

P. L. Dragotti, M. Vetterli, and T. Blu, Sampling moments and reconstructing signals of finite rate of innovation: Shannon meets strang-fix. *IEEE Trans. on Signal Process*, 55(5), Part 1: 1741–1757, 2007.

M. F. Duarte and Y. C. Eldar, Structured compressed sensing: From theory to applications. *IEEE Trans. Sig. Proc.*, 59(9): 4053–4085, September 2011.

M. Elad, *Sparse and Redundant Representations: From Theory to Applications in Signal*, Springer, New York, 2010.

Y. C. Eldar and G. Kutyniok (Eds.), *Compressed Sensing, Theory and Applications*, Cambridge University Press, Cambridge, UK, 2012.

J. J. Fuchs, On sparse representations in arbitrary redundant bases. *IEEE Trans. Inf. Theory*, 50(6): 1341–1344, June 2004.

R. Gribnoval and M. Nielsen, Sparse representation in unions of bases. *IEEE Trans. Inf. Theory*, 49(12): 3320–3325, December 2003.

S. Ji, Y. Xue, and L. Carin, Bayesian compressive sensing. *IEEE Trans. Signal Process,* 56(6): 2346–2356, June 2008.

L. W. Kang, C. Lu, and C. Hsu, Compressive sensing-based image hashing. In *Proc. Int. Conf. Image Process.,* pp. 1285–1289, 2009.

Y. C. Lin, D. Varodayan, and B. Girod, Distributed source coding authentication of images with contrast and brightness adjustment and affine warping. In *Proc. in Picture Coding Symp.*

S. Mallat, G. Davis, and Z. Zhang, Adaptive time-frequency decompositions. *SPIE J. Opt Eng.,* 33(7): 2183–2191, 1994.

S. Mallat and Z. Zhang, Matching pursuits with time-frequency dictionaries. *IEEE Trans. on Signal Process,* 41(12): 3397–3415, 1993.

M. Mishali, Y. C. Eldar, O. Dounaevsky, and E. Shoshan. Xampling: Analog to digital at sub-Nyquist rates. *IET Circuits, Devi. Syst.,* 5(1): 8–20, January 2011.

B. K. Natarajan, Sparse approximate solutions to linear systems. *SIAM J Computing,* 24: 227–234, 1995.

D. Needell and J. Tropp, CoSaMP: Iterative signal recovery from incomplete and inaccurate samples. *Appl. Comput. Harmo. Anal.,* 26(3): 301–321, 2009.

J. Nocedal and S. Wright, *Numerical Optimization,* Springer-Verlag, Heidelberg, 1999.

B. A. Olshausen and D. J. Field, Emergence of simple-cell receptive field properties by learning a sparse code for nature images. *Nature,* 381: 607–609, 1996.

J. Portilla, Image restoration through 10 analysis-based sparse optimization in tight frames. *Proc. IEEE Int. Conf. Image Proc.,* 2009.

L. I. Rudin, S. Osher, and E. Fatemi, Nonlinear total variation noise removal algorithm. *Physica D,* 60: 259–268, 1992.

C. E. Shannon, Communication in the presence of noise. *Proc. IRE,* 37:10–21, January 1949.

M. A. Sheikh and R. G. Baraniuk, Blind error free detection of transform domain watermarks. In *Proc. Int. Conf. Image Processing,* pp. 453–456, 2007.

H. R. Sheikh and A. C. Bovik, Image information and visual quality. *IEEE Trans. Image Process.,* 15(2): 430–444, 2006.

Q. Sun, Sparse approximation property and stable recovery of sparse signals from noisy measurements. *IEEE Trans. Signal Process,* 59(10): 5086–5090, 2011.

M. Tagliasacchi, G. Valenzise, and S. Tubaro, Localization of sparse image tampering via random projections. In *Proc. Int. Conf. Image Processing,* pp. 2092–2095, 2008.

M. Tagliasacchi, G. Valenzise, and S. Tubaro, Hash-based identification of sparse image tampering. *IEEE Trans. Image Process.,* 18(11): 2491–2504, 2009.

M. Tagliasacchi, G. Valenzise, S. Tubaro, G. Cancelli, and M. Barni, A compressive sensing based watermarking scheme for sparse image tampering identification. In *Proc. Int. Conf. Image Processing,* pp. 1265–1268, 2009.

A. M. Tillmann and M. E. Pfetsch, The computational complexity of the restricted isometry property, the nullspace property, and related concepts in compressed sensing. *Proceedings of Signal Processing with Adaptive Sparse Structured Representations,* July 8–11, 2013. EPFL, Lausanne.

J. A. Tropp, Greed is good: Algorithmic results for sparse approximation. *IEEE Trans. Inf. Theory,* 50(10): 2231–2242, October 2004.

J. A. Tropp, Just relax: Convex programming methods for identifying sparse signals in noise. *IEEE Trans. Inf. Theory,* (52)3: 1030–1051, March 2006.

J. A. Tropp and A. C. Gilbert, Signal recovery from partial information via orthogonal matching pursuit. *IEEE Trans. Inform. Theory,* 53(12): 4655–4666, 2006.

J. Tropp and S. Wright, Computational methods for sparse solution of linear inverse problems. *Proc. IEEE*, 98(6): 948–958, 2010.

S. A. Vavasis, *Nonlinear Optimization: Complexity Issues*, Oxford University Press, New York, 1991.

L. R. Welch, Lower bound on the maximum cross correlation of signals. *IEEE Trans. Inf. Theory*, 20(3): 397–399, May 1974.

X. Zhang, Z. Qian, Y. Ren, and G. Feng, Watermarking with flexible self-recovery quality based on compressive sensing and composite reconstruction. *IEEE Trans. Inf. Forensics Security*, 6(4): 1223–1232, 2011.

Compressive Sensing Solvers

1. http://users.ece.gatech.edu/~justin/l1magic/
2. http://sparselab.stanford.edu/
3. http://www.lx.it.pt/~mtf/GPSR/
4. http://www.stanford.edu/~boyd/l1_ls/
5. http://www.personal.soton.ac.uk/tb1m08/sparsify/sparsify.html
6. http://www.lx.it.pt/~mtf/SpaRSA/
7. https://sites.google.com/site/igorcarron2/cs#reconstruction (comprehensive listing of solvers)

Compressive Sensing Course Pages

1. http://www-stat.stanford.edu/~candes/stats330/index.shtml
2. http://www.ee.washington.edu/class/546/2010spr/
3. https://www.math.ucdavis.edu/~strohmer/courses/280CS/280CS.html
4. http://cnx.org/content/m18733/1.5/
5. http://cnx.org/content/col10458/latest/
6. http://thanglong.ece.jhu.edu/Course/648/
7. http://nuit-blanche.blogspot.in/(Comprehensive list)

Index